NEW PRODUCT DEVELOPMENT

An Empirical Study of the Effects of Innovation Strategy, Organization Learning, and Market Conditions

Sameer Kumar
University of St. Thomas
Minneapolis, Minnesota

and

Promma Phrommathed
University of St. Thomas
Minneapolis, Minnesota

Library of Congress Cataloging-in-Publication Data

Kumar, Sameer

New product development: an empirical study of the effects of innovation strategy, organization learning and market conditions/Sameer Kumar and Promma Phrommathed.

p. cm.

Includes bibliographical references and index.

ISBN 0-387-23271-0 — ISBN 0-387-23273-7 (e-book)

1. New products. 2. Strategy planning. 3. Orgamizational learning. 4. Competition. I. Phrommathed, Promma.

HF5415.153.K85 2005

658.5′75—dc22

2004061516

ISBN 0-387-23271-0 (HB) ISBN 0-387-23273-7 (eBook)

Printed on acid-free paper

Printed in the United States of America

9 8 7 6 5 4 3 2 1 SPIN 11327240

springeronline.com

This book is dedicated to Our Families, Parents and Friends

Preface

The challenge of managing a business enterprise today is to ensure that it can remain efficient and competitive in a dynamic marketplace characterized by high competition, unstable demands, heterogeneous market segments, and short product life cycles. Increasing the pace of new product introduction enables dealing with shorter product lives. To sustain competitiveness, a firm has to be innovative as well as quick to respond to the changing customer needs in order to provide better and faster products to market than competitors. New product development (NPD) is considered as a process of learning. Successful NPD projects typically rely on knowledge and experience of multi-function teams. In addition to corporate strategy and organization learning, the external factors such as, market and competitive conditions also play a big role in driving business strategies.

The results from the empirical research study reported shows that companies implementing innovation strategy are more competitive in the long run while those that follow customer-responsive strategy are more likely to have higher return on investment within a shorter time. In order to achieve both sustainable competencies and also meet customer needs in the changing market environment today, a company should adapt to the benefits of both strategies. Among other results from the study, the following are worth noting: (1) combining both innovative and customer-responsive strategies improve probability of product success when a new product is launched into a market; (2) organization learning and knowledge management can improve both NPD process and project success; and (3) the shortened product life cycles for companies require adjusting their market strategies by competing through product differentiation rather than lower price, and realizing time-to-market by reducing their product development times.

This book covers the subject of new product development and how it is affected by organization's innovation strategy, learning and market conditions through:

(1) an introduction to generic competitive strategies and new product development differentiation and also research questions triggering the study in Chapter 1;
(2) a basic conceptual model outlining types of new product development, organization learning and knowledge management's role in NPD, impact of market conditions on NPD strategies and total cost of ownership in a supply chain context in Chapter 2;
(3) research methodology used is outlined in Chapter 3;
(4) results and necessary analyses of the survey data in general terms are described in Chapter 4;
(5) results from the analyses of survey data specific to the propositions outlined in Chapter 2 are presented in Chapter 5; and
(6) relevant conclusions on the research postulates, limitations, recommendations for future research, examples of NPD in leading companies and illustrating technology function deployment approach for developing technology roadmap in Chapter 6.

The appendices include industry survey letter, questionnaire, and raw survey data.

Acknowledgments

We gratefully acknowledge all those who helped us in bringing out this book for publication. First and foremost we have greatly benefited from the wealth of a vast array of published materials on the subject of New Product Development, Diffusion of Innovation, Organization Learning, Market Turbulence and Supply Chain Management.

We would like to thank the reviewers of the manuscript of the book. The contents of this book has benefited immensely from their valued insights, comments, and suggestions.

The authors are grateful to the industry colleagues who participated in the "New Product Development" industry survey. Their painstaking efforts in responding to numerous questions in a fourteen pages long survey instrument were pivotal to developing this book. Names of participants are not listed here as it was a confidential survey.

Both authors owe gratitude to their families, parents and friends for their everlasting support. Finally, we wish to thank our editor, Mr. Ray O'Connell and the entire production team at Kluwer Academic Publishing, for their assistance and guidance in successful completion of this book.

Contents

Chapter #1

INTRODUCTION

If the world was stable and all supplies could just adequately satisfy all demand levels, corporations would have no need to compete with each other in the marketplace. Firms also would have no need to monitor and understand what has changed in the marketplace and what works well in business strategies in order to respond to such changes. In reality, the profit-seeking organizations function in dynamic environments- not stable ones, in order to gain profit, satisfy stakeholders, and even survive in extremely competitive business environment.

Both the internal and competitive environments in which the firms operate evolve over time. In response, management strategies must also change over time so that the firms can remain effective and competitive through the changing market situations (Griffin 1997). Competition is at the core of the success or failure of firms. It determines the appropriateness of a firm's activities that can contribute to its performance such as innovations, a cohesive culture, or a good implementation of its strategies. Two major corporate strategies leading to competitive advantage are *cost leadership* and *product differentiation* (Porter 1985)

1. GENERIC COMPETITIVE STRATEGIES

To be competitive through cost leadership, companies keep expanding to gain benefit from economies of scale in conjunction with exploitation of resources to improve process efficiency aiming to lower costs, maximal throughputs, and optimal resource utilization. Well known strategic and tactical tools used in conjunction with this strategy include Six-Sigma, Lean Manufacturing, Just-in-Time (JIT), Supply Chain Management (SCM), etc. Corporation pursuing this approach typically sell standard products with prices comparable to competitors. This practice may mislead the company to focus too much on costs and slash its quality to achieve the figures that in turn would ultimately push its prices down, too. This approach temporarily provides the successfully practicing companies an admirable outcome with respect to lower operating costs in the income statement, yet no opportunity for growth.

On the other side of an operating profit equation, exist the revenues or the sales as a result of products the firms have supplied to the markets. Another competitive strategy attempts to leverage the incomes by supplying a variety of products to successfully meet customer needs and expectations. The more product alternatives with reasonable prices firms provide to marketplace, the more likely they are to be financially successful. To excel through product differentiation, companies must implement market research, technology deployment, research and development (R&D), and innovation in products and processes, etc.

To compare and contrast these two strategies in terms of SWOT analysis, the cost leadership can be viewed as an internal factor that may be transformed into either strength (S) or weakness (W) of an organization. Any business concentrating in enhancing this process efficiency will turn this capability into strength. This capability unfortunately has limitation. The good firms will not squeeze efficiency beyond the point of providing returns that provide a favorable return on investments; the good firms will reduce costs incrementally without adversely impacting cost-service levels/relationships. That is the reason why this strategy has not much room for organization growth. On the other hand, the product differentiation can be seen as an external attribute that may be classified into either opportunity (O) or threat (T) for a business. This is the actual unlimited opportunity for a business to grow and stay competitive in market. Any firm that seizes this opportunity and turns it into strength will greatly enjoy the sustained fruits of investment.

2. NEW PRODUCT DEVELOPMENT DIFFERENTIATION

The strategies being debated in the business world and academia as mentioned above are the prime motive to pursue researching this aspect to discover the mechanics of successful new product development application. One of the significant tactics relating to product differentiation, that number of firms have implemented to enhance their competitive advantage, is efficient new product developmentand successful introduction of the new breed of products into markets. Key competitive advantage in marketplace is the ability to repeatedly commercialize successful new products (Griffin and Page 1996). In the U.S. alone, more than 10,000 new products enter the market each year (Crawford 1997). Unfortunately, these products often fail in the market or never enter the launch phase at all (Leenders, Wierenga 2002).

Today's business environment tends to be rapidly and non-linearly changing. Thus, business objectives are very much tied to management of unpredictability and chaos. Some of the changes in the market environment with the potential impact to the ways in which new product development is practiced and managed over the past decade include:

- Increased level of competitions in the same market
- Rapidly changing customer needs and expectations
- Higher rates of technical obsolescence
- Shorter product life cycles (Griffin 1997)

The primary impact of these environmental changes is to drive business organizations to implement those changes that help accelerate products through development and improve process efficiency as well as overall NPD effectiveness. Nevertheless, always moving along the changing environments may totally diverge the companies out of the planned pathway - away from original mission!

This research is therefore developed to study several aspects of new product development to be able to answer following questions:

- What are motivating factors encouraging companies to develop their new products?
- What is the silver bullet (best practices, or tools) used to accomplish the NPD goals?
- Do these tools really provide sustained competitive advantage to the exercising firms?

In order to answer these questions correctly, a research study was organized to collect data and information to support proposed hypotheses, which are discussed in detail in chapter 2- basic conceptual model. Chapter 3- research methodology discusses the approach taken for the research including literature survey and industrial survey intended to ratify the postulates. Chapter 4- results and analyses, portrays the summary of survey data accompanied by statistical analyses of the selected sets of data. The findings in chapter 5 focus on the verification of proposed hypotheses stated in chapter 2 with the analyzed data from chapter 4 and related previous researches. Finally, chapter 6- conclusions and ramifications summarizes the research outcomes and suggest further study opportunities.

Chapter 2

BASIC CONCEPTUAL MODEL

This chapter essentially discusses main themes of the research based on previous studies and research works carried out by scholars in academia and practitioners in industry. In addition, discussion on diffusion of innovation and total cost of ownership in a supply chain context is also provided. The research also takes practical data from an industrial survey targeting professionals associated with new product development in various industries. The procedure for establishing propositions in this study was different from most other researches reported in literature, which usually define propositions prior to developing survey instrument. This study started with developing a comprehensive survey covering as wide perspective as possible: - including not only such basic aspects that most of previous researches have focused on, such as, project management, customer involvement, R&D, development process, and financial measurement but also other NPD-associated aspects most researches had excluded; for example, corporate strategy, product success, leadership and teamwork, technology deployment, and market turbulence.

After getting survey feedback, information was extracted from the survey results that hopefully constituted knowledge for this field of study. The proposed hypotheses however have to be verified with surveyed results by means of analytical tools like analysis of variance (ANOVA) and regression analysis.

Eventually, there are three themes that can be synthesized from this research:- Typology of new product development, Organization learning and knowledge management's roles in NPD, and Impact of market conditions on NPD strategies.

1. THEME 1: TYPOLOGY OF NEW PRODUCT DEVELOPMENT

Before starting to identify types of innovation, it is important to begin with defining all relating terms to innovation that different scholars have used interchangeably in literature even if those terms have distinguishing meanings (Garcia, Calantone 2002). The most ambiguity might fall into the utilization of terms 'Innovation', 'Invention', 'Innovativeness', and 'Creativity' interchangeably in NPD literature. Next, the typology of technological innovation will be discussed. Finally, this study will propose a hypothesis of classification of NPD strategy based on corporate innovation strategy. The context of NPD typology can be illustrated according to figure 2-1.

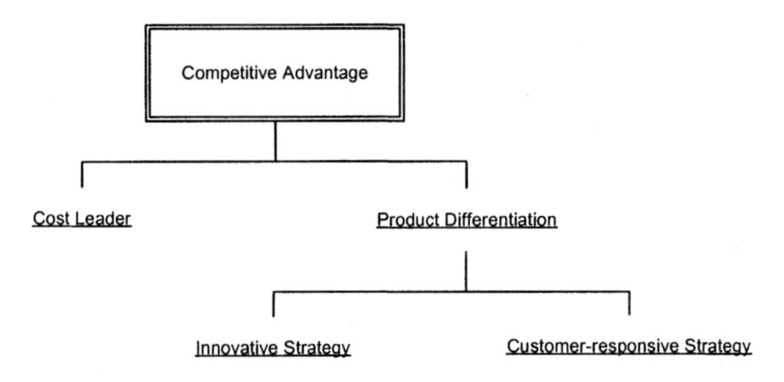

Figure 2-1. Context of innovation strategy and NPD strategy

1.1 Definition: Innovation, Invention, Innovativeness, and Creativity

'Innovation' is an iterative process initiated by the perception of a new market and/or new service opportunity for a technology-based invention, which leads to development, production, and marketing tasks striving for the commercial success of the invention. This definition addresses two important distinctions: [1] the 'innovation' process comprises the *technological development* of an invention combined with the *market* *introduction* of that invention to end-users through adoption and diffusion, and [2] the innovation process is *iterative* in nature and thus, automatically includes the first introduction of a new innovation and the reintroduction of an improved innovation.

'Invention' does not become an innovation until it has processed through production and marketing tasks and is diffused into the marketplace. Innovation includes not only basic and applied research but also product development, manufacturing, marketing, distribution, servicing, and later product adaptation and upgrading. A discovery that goes no further than the laboratory remains an invention. A discovery that moves from the lab into production and adds economic value to the firm (even if only cost savings) would be considered an innovation. Thus, an innovation differs from an invention in that it provides economic value and is diffused to other parties beyond the discoverers.

'Innovativeness' is most frequently used as a measure of the degree of *newness* of an innovation. 'Highly innovative' products are seen as having a high degree of newness and 'low innovative' products sit at the opposite extreme of the continuum. From a macro perspective, 'innovativeness' is the capacity of a new innovation to create a paradigm shift in the science and technology and/or market structure in an industry. From a micro perspective, 'innovativeness' is the capacity of a new innovation to influence a firm's existing marketing resources, technological resources, skills, knowledge, capabilities, or strategy.

'Creativity' is more likely referred to ability of organization to generate new ideas regardless of how much the ideas are developed into products. An organization may be so creative that it has number of ideas raised from executive and employees but not innovative since it does not transform the ideas into commercial products. "A creative thought is not worth anything unless you can translate that into something useful for the company", said William Coyne, senior vice president of R&D at 3M. That is why 3M prefers to use the word innovation rather than the fuzziness of creativity (Filipczak 1997).

1.2 Definition: Radical innovation, Incremental innovation, and Imitative innovation

To define categories of innovations, one may basically classify the technological innovations into 'Radical innovation' and 'Incremental innovation' based on the perception of newness of that innovation to

ology and market. However, another term of 'Imitative innovation' has also been misused by substituting 'Incremental innovation' from time to time. Here are the definitions of those terms defined by Garcia and Calantone 2000.

'Radical innovations' are defined as innovations that embody a new *technology* that results in a new *market* infrastructure. The radical innovations result in discontinuities on both a macro and micro level. Radical innovations often do not address a recognized demand but instead create a demand previously unrecognized by the consumer. This new demand cultivates new industries with new competitors, firms, distribution channels, and new marketing activities.

'Incremental innovation' can easily be defined as products that provide new features, benefits, or improvements to the *existing* technology in the *existing* market. An incremental new product involves the adaptation, refinement, and enhancement of existing products and/or production and delivery systems. Incremental innovations are important on two counts: first as a competitive weapon in a technologically mature market; and second, because streamlined procedures based on existing technology can help alert a business in good times to threats and opportunities associated with the shift to a new technological platform.

Grupp 1998 provides a very succinct definition of imitative innovations. Innovation occurs only in the first company to complete industrial R&D that culminates in the launch of the first product on the markets. Rival innovations are designated ***'Imitative innovations'*** even if, in intracorporate term, very similar R&D processes are only a short distance from one another chronologically. Because of their iterative nature, imitative products are frequently new to the firm, but not new to the market. In spite of high technological innovativeness, imitative innovations usually have low market innovativeness.

Most executives consider innovation as a key to competitive advantage. Accenture, a renowned global management and technology service firm, conducted a survey targeting 350 CEOs from a variety of industries on innovation strategy and implementation (Haapaniemi 2002). The study found that almost 60% of respondents said innovation is one of the five most important factors in building competitive advantage, and more than 10% said it is the single most important factor. The group of companies that valued innovation as the most crucial factor were mainly located in telecommunication, high technology, and multinational-operating industries.

1.3 DIFFUSION OF INNOVATION

The original diffusion research was done as early as 1903 by the French sociologist Gabriel Tarde who plotted the original S-shaped diffusion curve. Tarde's 1903 S-shaped curve still holds because most innovations have an S-shaped rate of adoption (Rogers 1983). The variance lies in the slope of the "S". Some new innovations diffuse rapidly creating a steep S-curve; other innovations have a slower rate of adoption, creating a more gradual slope of the S-curve. The rate of adoption, or diffusion rate has become an important area of research to sociologists, and more specifically, to advertisers and marketing professionals. In the 1940's, two sociologists, Bryce Ryan and Neal Gross published their seminal study of the diffusion of hybrid seed among Iowa farmers; renewing interest in the diffusion of innovation S-curve. The now infamous hybrid-corn study resulted in a renewed wave of research. The rate of adoption of the agricultural innovation followed an S-shaped normal curve when plotted on a cumulative basis over time. This rate of adoption curve was similar to the S-shaped diffusion curve graphed by Tarde forty years earlier.

In his 1995 comprehensive book *Diffusion of Innovations*, Everett Rogers defines diffusion as the process by which an innovation is communicated through certain channels over time among the members of a social system. Roger's definition contains the following four elements that are present in the diffusion of innovation process:

- *Innovation* – It is an idea, practices, or objects that is perceived as new by an individual or other unit of adoption.
- *Communication channels* – These are the means by which messages get from one individual to another.
- *Time* – The three *time* factors are:
  - *innovation*-decision process.
  - *relative time* with which an innovation is adopted by an individual or group.
  - *Innovation's rate of adoption.*
- *Social system* – It is a set of interrelated units that are engaged in joint problem solving to accomplish a common goal.

There are five stages (individual or group) of Innovation Diffusion:

Stage 1. Knowledge – exposure to its existence and understanding of its functions.

Stage 2. Persuasion – forming of a favorable attitude towards it.

Stage 3. Decision – commitment to its adoption.

Stage 4. Implementation – putting it to use.
Stage 5. Confirmation – reinforcement based on positive outcomes from it.

The Theory of Innovation Diffusion holds that there will be an increased rate of diffusion if potential adopters perceive the innovation:

- To have a relative advantage – the relative degree to which it is perceived to be better than what it supersedes.
- Is compatible – with existing values, past experiences and needs.
- It is not overly complex – ability to understand and use.
- Trial-able – the degree to which it can be experimented with on a limited basis.
- Offers observability – visible results.

All four elements (of diffusion of innovation process) play major roles in the diffusion of an innovation, how it will be accepted, and whether or not the innovation will survive. Another element to consider in the theory is Critical Mass. Critical Mass is the point in time in which enough individuals have adopted an innovation so that the date adoption becomes self-sustaining. It means that the innovation will survive. According to the concept of critical mass, an interactive medium is of little value unless other individuals also adopt the system. All of this may sound either complex or, then again, it may be fully understandable but to give light to how we see the acceptance or decline of new technologies or products, whatever the case, let us further explain with a short example.

Every year new technologies, from laptops and palm pilots to video games, hit the market (Innovation). These technologies are diffused through mass media be it print or electronic (Communication Channels). Some of these technologies will either be sold or be dismissed and many become the new "got to have it" item in a matter of weeks (Time). The consumer in this instance, are those at the receiving end, those that purchase the technology (Social System).

Diffusion theory has often been used to model the first-purchase sales growth of a new product over time and space. Diffusion theory suggests that there is a time lag in the adoption of products by different members of a social system. A select few innovators who, in turn, influence others to adopt it first adopt the product. Thus it is the interaction or interpersonal communication (word-of-mouth) between adopters and non-adopters that is posited to account for the rapid growth stage in the diffusion process (Rogers and Shoemaker 1971). The best known first-purchase diffusion models of new product acceptance in marketing are those of Bass (1969), Fourt and Woodlock (1960), and Mansfield (1961). In particular, the Bass model has been successfully demonstrated in retail service, industrial technology, agriculture, and consumer durable sectors.

It is established that the introduction of an innovation may affect the diffusion process of another innovation, provided the two are sufficiently related by function or application (Alpert 1994). For example, Peterson and Mahajan (1978) developed diffusion models for complementary innovations. However, the Peterson and Mahajan approach anticipates a direct interdependency between the innovations, such as the relationship between computing hardware and software. In contrast, Redmond (2002) study focused on the propensity of communication innovations to accelerate the diffusion of indirectly related or unrelated innovations. Redmond examined the role of market competition in fostering the diffusion of communication innovations and thus of impacting the diffusion of innovations generally. The broad-scale impacts of this interaction include a shift in the social evaluation of newness and a shortening of product life cycles.

3. INNOVATIVE VERSUS CUSTOMER-RESPONSIVE STRATEGY FOR NPD

It is not atypical for firms to develop new products by following two main principles: - innovative strategy or customer-responsive strategy. The ***Innovative*** strategy basically focuses on radical and incremental innovations while the ***Customer-responsive*** strategy emphasizes market relation as the input to NPD process. Companies pursuing the innovative NPD strategy can be called as *Prospectors* who value being first with new products, markets, and technologies even though not all efforts prove to be profitable. The other group can be called as *Reactors* who seldom are first to market with new products but rather carefully listen to customer voice and respond to it with more cost-effective products (Griffin, Page 1996). For repeated number of times, the reactors may come up to the market with new products that better reflect customer needs than those of the prospectors. That is one of the ironic reasons that discourage firms to pursue innovation.

Among the innovative strategy advocates, Peter Drucker, Peter Senge, and Gary Hamel believed that only radical innovation could lead to significant growth (Drucker 1998; Senge 1998; Hamel 2003). True innovation is based on the recognition that a business concept represents a number of design variables, all of which need to be constantly revisited and constantly challenged. Dell, Starbucks, and Wal-mart, as good examples of innovations, developed a new business model based on changing technology, demographics, and consumer habits. To be truly innovative, firms have to cultivate the concept into employees' minds through mission, vision, and values. Nokia is another company that established an innovation-supporting environment. Its strategies does not only come from top management, but they stem from hundred of ideas employees created to focus on how to *humanize* technology. The real work of top management is not to generate the new thinking but to look at all these ideas and try to find the fundamental themes that would give overall direction to the company. To date, Nokia has expanded its capabilities and its virtual presence much faster than anybody else in the same industry (Hamel 2003).

On the other hand, the creation of successful new products is fundamentally a multidisciplinary process. Consequently, firms are embracing coordination mechanisms e.g. Quality Function Deployment (QFD) and cross-functional teams, aimed at increasing the level of not only internal interaction but also customer integration (Olson et al 2001). A study found out that a careful investigation of customer needs and best responding to such needs could improve customer satisfaction; and in turn could promote the company's profits. A study comparing U.S. and New Zealand small entrepreneurial firms proved that, the reason that the New Zealand firms had higher level of NPD performance than those of their U.S. counterparts was the greater emphasis on marketing and customer-focused NPD practices (Souder et al 1997).

The following table summarizes attributes of innovative and customer-responsive strategies for NPD.

Table 2-1. Summary of characteristics of prospectors and reactors

Innovative Strategy (Prospector)	***Customer-responsive Strategy (Reactor)***
• *Technology (product) driven*	• *Market (customer) driven*
• *Push to market (High risk when commercialized)*	• *Pull from market*
• *Producer dictates pricing*	• *Consumer dictates pricing*
• *High entry barrier, low competition*	• *Low entry barrier, high competition*
• *Radical + Incremental innovation*	• *Incremental + Imitative innovation*
• *Back-office activities*	• *Front-desk activities*
• *Highest returns in downward economy*	• *Highest returns in growing economy*

3.1 Innovative Firms (Prospectors):

Based on the attribute summary in table 2-1, prospectors try to excel with innovation by being the first to market and maintaining leading position of market share and sales volume. Their key competitive advantage is achieved through technology advancement that is hard for competitor to follow and overrun. The prospectors vastly invest in R&D, design, and manufacturing activities reflecting processes that contribute to technological advancement. Concentrating only at the intrinsic processes to leverage technical competency often lures employees to be overly proud of their successes without considering surrounding entities e.g. customers, suppliers, and competitors. This adversely leads such a company to purely push its product innovations into market regardless of whether customers want them or not. This practice turns to hurt the company itself since the new products may not be successfully commercialized. Nevertheless, innovation does not merely means being the first product in the market, but it also covers innovations in services and marketing that can expand market opportunity.

Being a lone leader in new product market automatically implies ability to specify product prices. In general, customers are willing to pay for any price for innovative products. This group of consumers perceives the innovative features as their prestige to process the products. Another favorable point of being innovative is that the pursuing companies will enjoy their profit over a high wall of entry barrier. It is not easy to trespass into a high technology battlefield unless the offenders are also equipped with high technology gears. That is the reason why there is usually low competition in the high technology industries.

Since the strategy focuses on being a leader in innovation, innovative firms can generate various types of innovations including breakthrough: - totally new to the firm and market, and incremental innovations that add features to existing products. Both types of innovations are the results of in-house development that effectively adds knowledge and experiences to organization's competency. In the purely innovative organization, top management pay most attention on back-office activities that drive technology and innovations like research and development, engineering, design, and manufacturing. Prospectors tend to focus less on front desk where customer and supplier relations take place.

By considering organization's surviving capability over economic conditions, prospectors can survive in any economic situations even in the hard times. In economic recession, only innovative firms can fully utilize their competency to strive for most profits out of the limited market opportunity. However, during good economic times, the prospectors can also gain profits but not as much as the opponents because the reactors can make more and faster profits with less cost.

One of the great innovative companies is 3M whose corporate strategy is innovation. Innovation is so important for 3M that it is clearly stated in the company's vision "To be the most innovative company in the markets it serves". To promote that innovative spirit 3M has been driven by the following set of strategies (Conceição et al 2002).

30%/4 Rule: This rule states that each year 30% of 3M's sales must result from products introduced in the last 4 years.

15% Rule: One of the most important principles in management at 3M is the "15% rule", which allows technical people to spend 15% of their time in projects of their own choosing without needing approvals or even without having to tell management in what they are working on.

Products are a division's responsibility: Each division is oriented directly towards the market, aiming at responding to its customer base.

Technologies are shared throughout the company: While products' responsibility lies within each division, technologies belong to the company. Every division has access to the technology resources of the entire company but also has the responsibility to share technological needs of its customers throughout 3M.

Technology combinations: 3M promotes the combination of its core technologies in order to realize advantages of those synergies to innovate new products and new applications.

Strong intellectual property protection.

3.2 Customer-responsive Firms (Reactor):

Reactors, on the contrary, try to capture what their customers' want and quickly respond to the needs with lower costs and higher efficiency of new product development process. As a result, dominating activities in organization involve customer service, sales, and market research. New product ideas are generally raised from customer needs. With this approach, products are pulled from market, which can partly guarantee success in the market.

However, since the customers know the trends of new products, they can control the prices. In addition, there might be more producers than just one company being asked by the customers to develop the same products to supply the same market. This approach typically leads to high competition because the entry barrier is low: - new entrants can get into the battlefield easily without high wall of technological barrier to adopt. Types of innovations relating to this strategy may be only incremental and imitative innovations because the implementing companies do not have much of their own knowledge and resources to develop in-house technology. Consequently, they have to rely on adopting technology or adapting pioneer's ideas to develop their new products.

The reactors typically gain highest profits during growing economic duration when consumers are not much selective. In addition, having low cost from economies of scale, high production efficiency, and quality that exactly matches customer needs really helps the firms to leverage their profit margins.

4. SWOT ANALYSIS OF NPD STRATEGIES

The analysis of strengths, weaknesses, opportunities, and threats, also known as SWOT analysis, is widely used in both academic and business world to compare and contrast two or more entities. The tool covers not only strengths and weaknesses, which are the internal attributes of the compared items but also such external characteristics as opportunities that might add into their strengths and threats that may turn into weaknesses. The study is to provide a better insight into innovative and customer-responsive strategies through this analyzing tool. The summary of SWOT analysis is depicted in table 2-2.

Table 2-2. Summary of SWOT analyses of NPD strategies

Innovative Strategy (Prospector)	*Customer-responsive (Reactor)*
Strengths • *Technology-driven: hard to imitate* • *Long-term growth: continual competitive advantage* • *Knowledge + skills turns into organization competency* • *High benefit for being first-to-market: price control, feature preference control* • *Increase customer loyalty*	***Strengths*** • *Market-driven: customers need it, lower market risks* • *Study from pioneer's failure: higher rate of success* • *Quick returns on investment* • *Benefit from higher efficiency*
Weaknesses • *Higher risk of market unacceptance* • *High investment: technology, resources* • *Slow returns on investment*	***Weaknesses*** • *New market entrants need more efforts* • *Short-term profit: not sustained* • *Easy to imitate by other new entrants: leads to high competition* • *Customers dictate prices*
Opportunities • *In economic crisis, innovative firms survive* • *Innovative firms attract innovative employees* • *Patent + licensing help protect competitive advantage*	***Opportunities*** • *In growing economy, higher profit* • *The strong financial condition draws investment* • *Technology can be easily acquired*
Threats • *How to retain knowledgeable employees?* • *Trends of customization reduce innovation needs* • *Slow returns may discourage investors*	***Threats*** • *Growing competition: lower margin* • *Advance technologies help companies to imitate one another*

Prospectors' SWOT:

Strengths:

- Being driven by technology, prospectors are prevailing in the competition since the competitors will find difficult to acquire or copy the same technology.
- Innovativeness is a vital asset of a firm that can give a long-term, sustained growth since such an asset can continually generate new products over time.
- The more innovation is practiced, the more knowledge and experiences the firm gains. Finally, such knowledge turns to be corporate competency that is hardly imitated by rivals.

- Prospectors typically become first-to-market that favors the market leader through abilities to control product prices, features, and cost of switching to other brands.
- Most customers usually become loyal to the innovative companies because they are confident that the firms would always introduce innovations to the market and they are willing to pay any prices for such products.

Weaknesses:

- Not all innovations are successful in market. Only innovative features without real demand in market can turn the bright dream down, too. A successful new product launch comprises a right product with a right price for a right market at a right time.
- Innovative organizations usually spend major investment in developing and acquiring technology and resources leading to low profit margin in the early years.
- Being an innovative firm needs very strong financial support because the return on innovation investment takes a relatively long time. In some industries it may take 10-20 years to realize first profit.

Opportunities:

- Innovative firms can grow or at least survive in any economic conditions. They can even benefit more in a hard time than another group does. For instance during the most recent US recession, only highly innovative companies like Microsoft, Dell, 3M, and FedEx that not only enjoyed profits from operation but also invested more in technology and resources--waiting to gracefully soar when the economy picks up.
- Most innovative workers prefer to join companies that promote creativity and innovation because they can utilize their talents freely. This would in turn benefit the companies.
- Current innovation protecting laws and regulations help prolong profit delight for innovative firms.

Threats:

- Amidst the merger and acquisition trends today, it becomes highly challenging on how to retain knowledgeable human resource from increasing relocation of workforce.
- Mass customization is challenging innovative firms by providing variety options of products with more cost effectiveness. However, the prospectors can easily copy this technique and become dominant over their precedents, too.

Reactors' SWOT:

Strengths:

- With market-driven approach, the reactors tend to be successful in commercializing new products at the first step because they develop the products that market have wanted. Marketing risk turns to be lowest with this practice.
- As fast-followers, the reactors enter market with higher rate of success because they have learned all successes and mistakes experienced by the pioneers. That is the reason why the most followers are relatively more successful than their precedents.
- With less investment on innovations and technology and faster product development time, the reactors tend to gain returns on investment quickly.
- After realizing mistakes the pioneers have experienced, the reactors typically design their process well to achieve higher efficiency in producing products.

Weaknesses:

- In market environment where market leaders have established high entry barriers such as technology level or cost of switching brand, the reactors can find themselves very difficult to enter and compete with such leaders.
- On the other hand, if the market is easy to enter, there would be number of new entrants getting in to fiercely compete with each other. The reactors eventually enjoy only a short-term profit. If they were not constantly moving forward, they would fall into the luring quick sand.
- Since customers know every movement of next-generation products, reactors have no other choice but to accept the prices indicated by the market. Even worse, if the customers also discuss the new product ideas with other suppliers, the reactors may be surprised to see the same products out of opponent's camps on the new product-launching day. Eventually, the consumers will benefit from this market.

Opportunities:

- During a strong economic situation, the reactors tend to gain higher profits than the prospectors due to their substantially lower costs.
- The admired financial figures of reactor, even if they are just a short-term result, can easily draw attention of investors, especially the short-term ones. This practice can help the reactors to expand quickly.

- The information technologies like product life cycle management system (PLM), customer relation management system (CRM), and product data management system (PDM), that help effectively convert customer needs into product design data, is ubiquitous today. The reactors can acquire as much technology as they can afford. That is an example of utilization of technology to shorten the time to market.

Threats:

- The competitive advantage of reactors is not really sustainable because it is easy to imitate by competitors. As a result, the competition will become intense and finally bring the profit margin down.
- The ubiquitous technologies that help companies compete with each other are like a double-edged sword - easy to acquire as well as easy to be copied.

5. MIDDLE LANE TO EXCELLENCE

With common sense, nobody would fully advocate only one extreme. By the same token, no company would only stick to one single policy- either fully innovative or fully customer-responsive, since the weaknesses of each are too failure-prone to bind company's destiny to the single one. Successful companies are organizations that value employee creativity, support innovations, as well as involve customers in NPD process. Arming the organization with innovation in conjunction to shielding the business success with customer involvement lead the exercising companies to excellence in operations.

To validate this assumption, a hypothesis is proposed and empirically verified as follows.

Proposition 1: *Companies pursuing both innovative and customer-responsive strategies for NPD tend to be more successful in business than companies engaging in only either one approach.*

The way to measure successes in business of introducing a new product into market should not simply focus only on financial figures such as revenues, profits, and market share. Other significant measures may include level of customer satisfaction and innovation perception of customers. To verify the hypothesis, the survey results relating to NPD strategies were used as the independent variables and compared with the dependent variables relevant to product success.

6. THEME 2: ROLES OF ORGANIZATION LEARNING AND KNOWLEDGE MANAGEMENT IN NPD

Due to the fact that today's business is becoming more complex, dynamic, and globally competitive than that of the past eras, business-related knowledge has been identified as the most important factor ultimately defining organization competency (Kock et al 1997). The speeding up of new product design and delivery, combined with the emergence of automation technology, has led to an increase in demand for knowledge workers and an increase in the knowledge-intensive labor component of products. At the same time, the amount of physical labor has also substantially reduced by the same means. According to Drucker's research (Drucker 1996), in 1880, about nine out of ten workers used their energy for physical works like making and moving things. Today, such a labor ratio is down to one out of five. The other four out of five are knowledge people or service workers. These workers converse on the phone, write reports, attend meetings, and so on.

Despite such a dramatic change, corporations in the modern days do not realize and adjust to change. One explanation can be that most institutions are deeply overwhelmed with the only one pole of Douglas McGregor's Human Side of Enterprise: Theory X rather than Theory Y. According to the theory X, employees are unreliable and uncommitted -- they just work for the paychecks. There would be no way such employees can contribute to corporate knowledge and competency. Most people treat the business enterprise as a machine for making money rather than as a living community. Consequently, employees are viewed as 'human resources' waiting to be employed (or unemployed) to the organization's needs (Senge 1998). Based on this presumption, executives tend to not lead nor support the learning habit of people and systems of the organizations, but rather push ideas down from the top level and let all others follow.

In the present turbulent environment, however, it is difficult to sustain such a top-down approach with the rest of the organization following in the grand master's footsteps. Learning at all levels of the organization has become increasingly important, as organizations have become flatter in order to cope with uncertainty. When things become uncertain and too complex, people tend to go back to the roots with the hope that a re-examination of the basics will afford some light on the present difficulties. The high degree of uncertainties over the last twenty years has caused organizations to re-think how they can use human abilities in order to exert some influence and control over such turbulence. Learning is a very basic human ability, a skill we all possess intrinsically and continue to use throughout our lives. What are gurus talking about learning and knowledge, then?

7. LEARNING AND KNOWLEDGE

Peter Senge has defined the meaning of **Learning Organization** in terms of continuous development of knowledge and capacity that "Learning organization is an *organizations where people continually expand their capacity to create the results they truly desire, where new and expansive patterns of thinking are nurtured, where collective aspiration is set free, and where people are continually learning how to learn together"* (Senge (1) 1990). Following Senge, Richard Karash defined terms related to knowledge and learning in a much more simple way that **Knowledge** is a capacity for effective action and **Learning** is an increasing knowledge that is in turn increasing capacity for effective action. In aggregate, **Learning Organization** is established when the organization and its people are continuously increasing their capacity to produce the results they really want to produce (Karash 1995).

Consider another term of **Organization Learning**[1] (OL) that this research study intended to specifically use in this study. John Shibley has defined the term that OL is a result of collective activities acted, reflected, and repeated by people in the organization (Shibley 2001). The knowledge and experiences from the former actions, after being reviewed and studied, will serve as the model for the next actions. A simple model of OL according to Shibley can be depicted as in the figure 2-2.

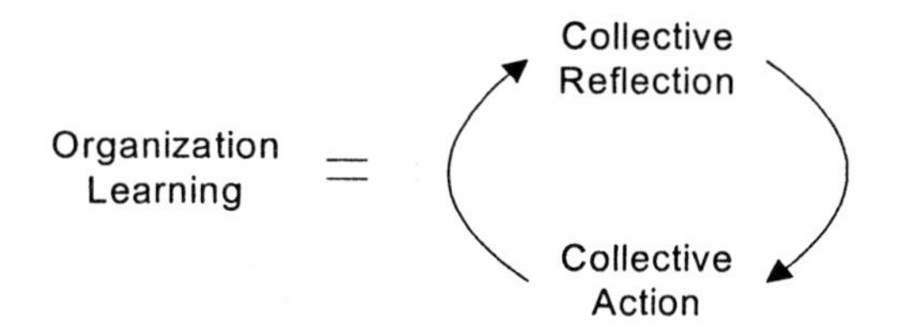

Figure 2-2. A basic model of organization learning as a result of repeating actions and reflections in organization

[1] Note that it is different from the previous term 'Learning Organization' that Senge and Karash had defined

Knowledge cannot be effectively obtained without considering its media: data and information. The concepts of data, information, and knowledge are closely related in spite of their distinct meanings. As commonplace as the confusion of data and information, is the confusion of knowledge and information.

Data is a carrier of knowledge and information, a means through which knowledge and information can be stored and transferred. Both information and knowledge are communicated through data, and by means of data storage and transfer devices and systems. In this sense, a piece of data only becomes information or knowledge when its receiver interprets it. On the other hand, information and knowledge held by a person can only be communicated to another person after they are encoded to data. Printed out paper and computer disks are examples of data storage devices. A corporate e-mail and the international airmail systems are examples of data storage and transfer systems.

While ***Information*** is *descriptive*, that is, it relates to the past and the present, ***Knowledge*** is eminently *predictive*, that is, it provides the basis for the prediction of the future with a certain degree of certainty based on information about the past and the present (Kock et al 1997). From this perspective, statements of the type "The water boiling temperature is at 100 degree Celsius at the normal atmospheric pressure" convey information, whereas statements of the type "If water is boiled at the top of a mountain, the boiling temperature will be under 100 degree Celsius" convey knowledge. Both knowledge and information flow between functions in organizations in the form of data transfer by means of media such as records, instructions, drawings, minutes of meeting, or even electronic forms like files and web pages, etc.

8. DATA, INFORMATION, AND KNOWLEDGE MANAGEMENT SYSTEMS

Figure 2-3 illustrates flowing streams of data from all phases of an NPD cycle to a database on real-time basis.

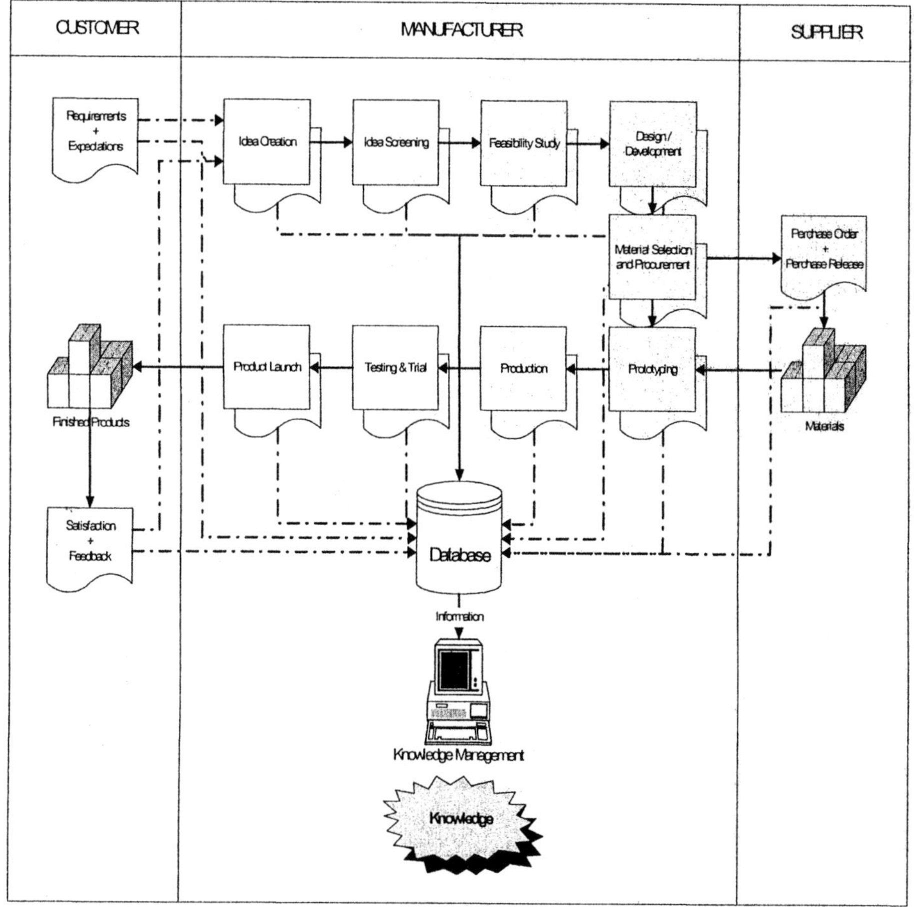

Figure 2-3. A model for data, information, and knowledge flows in an NPD cycle

The database management system, residing in the database, will then rearrange the data into predefined categories, which will be easy to organize and retrieve in the future. At this stage the scattering data - in the form of

Online Transaction Processing (OLTP) data - are transformed to information, sometimes called Online Analytical Processing (OLAP) data, which help create multi-dimensional data views. Although the information obtained from the database can give advantageous decision-supports but it relies very much on the person who inputs queries to retrieve such information via a standard language called Standard Query Language (SQL). Moreover, the information exhibits only past description but cannot generalize into future knowledge. As a result, knowledge management systems are developed to synthesize knowledge from collective information.

8.1 Data Mining

There have been numbers of systems built for this application. One of the widely applicable systems is Data Mining that has its own intelligence to discover unsuspected patterns from vast data warehouse. The most commonly used algorithms in data mining are neural networks, decision trees, genetic algorithms, nearest neighbor, and rule induction (Thearling 2002).

Some of these techniques have been used in specialized analysis tools that work with relatively small volume of data. These capabilities are now evolving to integrate directly with industry-standard data warehouse and OLAP platform. At the moment, data mining widely provides next generation decision-support services for various applications; such as, Internet e-Commerce, direct marketing, healthcare, genetics, customer relationship management (CRM), and financial services.

8.2 Product Lifecycle Management (PLM)

While data mining, using intelligent algorithms to discover hidden correlations of data, may be broadly used in a variety of business units, PLM is more commonly used in manufacturing enterprises. The PLM Center of University of Michigan describes Product Lifecycle Management (PLM) as an integrated, product-oriented, information-driven approach to all stages of a product's life – from its design inception, through its manufacture, development and maintenance, and culminating in its removal from service and disposal. The system lets the different groups exchange and collaborate on product-related images and data in real time with goals to improve manufacturing efficiency, product quality, and time-to-market (Vijayan 2003). PLM thus involves everyone in the manufacturing enterprise, including engineering, manufacturing, logistics, sales, marketing, and customer service.

PLM can be viewed as an extension of organization learning and lean manufacturing. It is an effort to use collective product-related information to help companies minimize the use of their resources by not wasting them in producing wrong or faulty products. PLM assumes that the data to provide this information come from the entire product's life and not simply from one phase. What adds to power, complexity, and difficulty of PLM is that it also extends collaborative engineering, product data management (PDM), enterprise resource planning (ERP), customer relationship management (CRM), and most other manufacturing systems (Grieves 2003).

9. HOW TO IMPLEMENT ORGANIZATION LEARNING?

Becoming a learning organization will involve a tremendous amount of attention, energy, and changes. Though, everyone realizes that both organizations and employees gain mutual benefits from this learning. The organization learning provides win-win prospects for the organization itself that achieves higher level of business performance and also the people who find personal reward and satisfaction (Karash 1995).

A wonderful story sums up the power of learning. It comes from George Fisher, former chief executive of Motorola and now chief executive of Kodak. 'Compare the rate of learning and improvement in the semiconductor business with the rate of learning and improvement in the automobile industry,' he said. 'If the automobile industry had learned at the same rate as semiconductors, a Cadillac which 20 years ago cost $10,000 would today cost 80 cents. That's the good news. The bad news is that it would be two inches long.' (Garvin 1997)

Peter Senge (Senge (1) 1990) has identified five disciplines essential to constitute the organization learning. These are disciplines in the sense that they deserve study and attention for a very long period of time, perhaps throughout a lifetime, not skills to be mastered quickly. The disciplines consist of:

- *Personal mastering*: Developing the capacity to continuously clarify what is important to an individual and how to achieve it.
- *Mental Modeling*: Developing the capacity to reflect on internal pictures of the world to shape one's thinking and actions.

- *Shared Vision*: The ability of an organization to create a deeply meaningful and broadly held common sense of direction.
- *Team Learning*: Developing capacity for collective intelligence and productive conversation.
- *System Thinking*: The ability to see the whole, perspective long-term patterns, understand interdependencies, and better recognize the consequences of actions.

As another perspective, David Garvin, a Cizik Professor of Business Administration at Harvard Business School, has defined the learning mechanism into two parts or skills. The first is creating, acquiring, and transferring knowledge; the second is modifying behaviors to reflect new knowledge and insights. Intel and General Motors were raised as examples of expensive lessons both received after they failed to realize these two skills respectively (Garvin 1997). In addition, Garvin also proposed five important characteristics of learning organization as follows.

- Systematic problem solving
- Experimentation with new approaches
- Learning from own experiences and past history
- Learning from best practices from competitors and world leaders (benchmarking)
- Transferring knowledge throughout the organization

An organization is more successful if its employees learn quicker, and implement and commercialize knowledge faster than the workers of the competition. An organization that does not learn continuously and is not able to continuously list, develop, share, mobilize, cultivate, put into practice, review, and spread knowledge will not be able to compete effectively. That is why the ability of an organization to improve existing skills and acquire new ones forms its most tenable competitive advantage.

To validate this assumption, a hypothesis is projected and to be empirically verified in chapter 5.

***Proposition 2**: NPD process and project successes base on implementations of organization learning and knowledge management.*

The ratio of the cycle time difference over the cycle time of last 10 years, which determines percent of NPD cycle time reduction, was used as the process-related success factor in this study. The project on time ranking that asked the respondents to rank ability of their companies on completing NPD projects on time was chosen to represent the project-related success.

For the independent variables, results from questions relating to organization learning and knowledge management were selected for study. Since there were two dependent variables, the proposition 2 can be split into two postulates as follows.

Proposition 2a: *NPD process success positively relies on implementation of organization learning and knowledge management.*

Proposition 2b: *NPD project success positively relies on implementation of organization learning and knowledge management.*

Both hypotheses were validated with the survey data. The results are discussed in details in the chapter 5.

10. THEME 3: IMPACT OF MARKET CONDITIONS ON NPD STRATEGIES

Enterprises in all branches of industry are being forced to react to the growing individualization of demand, yet, at the same time, increasing competitive pressure dictates that costs must also continue to decrease. Companies have to adopt strategies that embrace both cost efficiency and a closer reaction to customer needs (Piller, Moeslein 2002).

10.1 Market turbulence

In the 21st century, enterprises situate in a new era of the evolution of business operation and structure. From the producer's point of view, market conditions at the present day are highly unpredictable and fluctuating in terms of volume and quality of both products and services. Such instability of customer demands is the consequence of globalization and technology advancement. Consumers have more choices on selecting products supplied from the local makers as well as overseas manufacturers, who might take opportunity of lower production costs to be competitive in foreign markets. Furthermore, information and communication technologies (ICT) are facilitating people around the world to communicate fluently and borderlessly.

To cope with these challenges, manufacturers have to adjust themselves to be more reactive to changes in the markets. Traditional mass markets are becoming more and more heterogeneous. Market segments scale down to niches and micro niches, where producers have to offer customized products or services for each individual customer. Thus, the supply constantly changes from mass production to customization process. A Delphi-Study of the German Federal Ministry of Education and Research found that mass market and today's common segmentation of markets and the related systems of variant manufacturing will be substituted by an individualized reworking of the market related to the individual customer (EuroShoe 2002).

10.2 Drivers of individualization

Everyone realizes that business-to-business (B2B) markets have always been characterized by customized products and services. A supplier in the B2B markets gains competitive advantage by providing goods and services that fulfill this individual requirement. In business-to-consumer (B2C) markets, however, new changes in business landscape lead to growing customization. Several reasons can explain this observation.

- The desire for customized products is growing along with increasing wealth (higher incomes, more leisure time, and higher level of education).
- Due to increasing level of education and the lifelong learning process, more and more people recognize complexity of problems and learn to encounter in different perspectives, which in turn, transform to private shopping habits.
- People are becoming more *hedonistic*, which means customers no longer want to "have" because one has, but rather "be and experience". Additionally, they are changing to be more demanding and choosier, whereas their behavior is becoming more spontaneous and less predictable (EuroShoe 2002).
- The saturation of demands in markets causes a high level of competition. Supplies have to be shifted from established markets of existing products to new products. To achieve strategic competitiveness, the new products have to better meet or exceed customer's needs and expectations with regard to reasonable prices and acceptable quality.

The EuroShoe Consortium summarizes the transition of demands and supplies over time as shown in Figure 2-4.

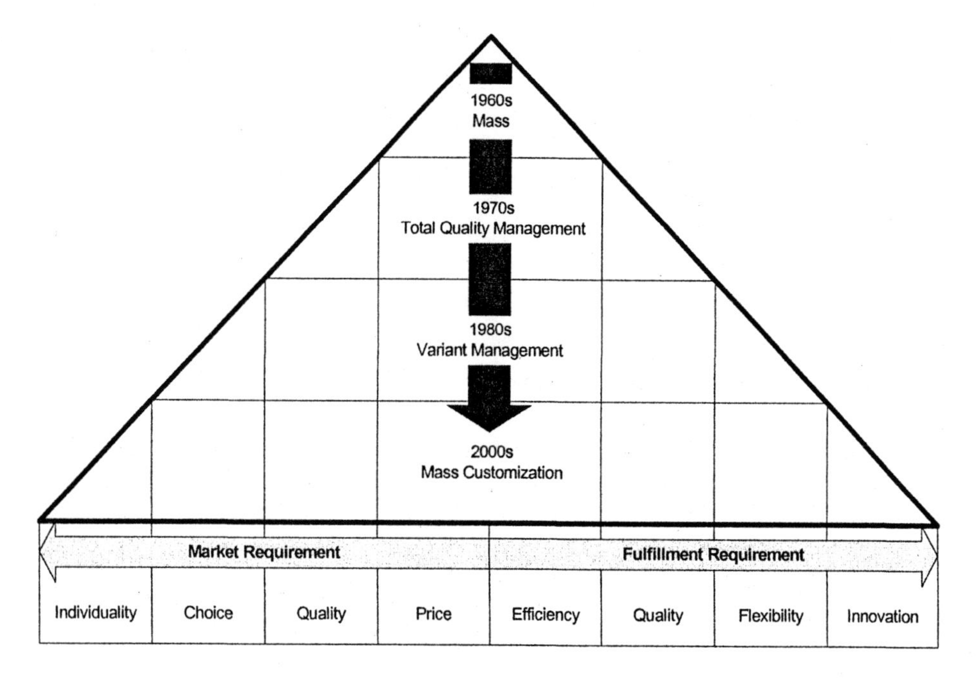

Figure 2-4. Revolution of demands and supplies from mass marketing and customization

Starting from the era of industrial revolution, markets in 1960s were characterized by pricing focus while manufacturers responded to the demands with efficient mass production to bring cost down. In 1970s, quality issues prevailed over the price and efficiency. However, it did not mean the low price and production efficiency were swept away but remained as basis besides the dominating quality hotshots. Total Quality Management (TQM) was introduced into industries to enhance quality as well as cost and efficiency. A decade later, consumers expected more variety of products. As a result, producers employed variant management to offer flexibility in products and related processes. In the present day, consumption demands more in term of individuality on top of existing reasonable price, good quality, and flexibility of products. The production can respond to the uniqueness customers want with innovation of its products.

Mass customization meets this challenge by producing goods and services to meet individual customer's needs with near mass production efficiency. The competitive advantage of mass customization is based on combining the efficiency of mass production with the differentiation possibilities of customization. From a strategic point of view (as discussed in the theme 1), mass customization means differentiation through customization, leading to greater product attractiveness. From the perspective of fulfillment, customization means the production of goods in so many variants that the needs of each relevant customer are fulfilled.

11. PRINCIPLES OF MASS CUSTOMIZATION

Information and efficient information handling can be regarded as the most important factors for the implementation of mass customization (Li, Possel-Doelken 2001). Contrary to the traditional made-to-stock system of mass production, the mass customized production pull the products along the entire production process based on the information of customer wishes. The information has to be translated into concrete specifications and has to be communicated from the points of sale back to the manufacturing and other fulfillment operations. In addition to information and information handling systems, product fulfillment systems are based on some other distinctive principles, which are complementary and hold strong synergy potential. The principles of mass customization can be described as follows.

- *Modularization of product architecture*: Following the concept of reusability and a product family design approach, most mass customization systems are based on a strong modular manufacturing system. The pre-fabrication and pre-construction of different parts facilitates the use of economies of scale and specialization, while an assemble-by-order system combines the customer specific selection into concrete products. Dell is probably the best example of mass customization implementation; customers can specify specifications of their own computer online. Once the order is confirmed, the assembly process can start within 24 hours (Brady et al 2000; Chanover 2000; Mello 2001; Torbenson 1998).
- *Limitation of customization possibilities:* Mass Customization does not aim to offer everything, or satisfy every single customer's wish. It does offer pre-designed options of product features that are hardly available in mass products. A consumer can thus select to customize the features into a single product. Levi Strauss does not offer truly tailor-made jeans that fit everyone, but a selection of 420 pre-defined sized plus the choice of a limited range of design options for each size (EuroShoe 2002).
- *Modularization of processes and stable fulfillment processes*: Due to the modular product architectures and the customization options, the mass customization manufacturers have a big challenge to design appropriate

production processes to fulfill two contradictory requirements. The production processes may start with mass production of modularized components. After the customized orders are placed, the final assembly processes are then customized to the various orders.

- *Build-to-order approach*: Mass customization production advocates lean manufacturing and does not stock up finished products. The inventory in this approach is therefore in the form of raw materials and pre-fabricated works-in-process (Anderson 2002).
- *Use of dedicated information systems*: The information systems are necessary for product specification configuration, production planning, order tracking, and relationship management. The role of information technology can be seen on two levels. First is the substitution of product functions by information activities, for example, order taking function is now perfectly replaced by online shopping web sites. Another role is data handling and management. Data of all transactions are not only processed through the ordering and payment systems, but also further analyzed to develop into knowledge that can lead to better prediction of the front-end mass production processes. Besides Dell's in-house developed order-handling system, Gerber Scientific Inc. has developed widely used systems that help its clients customize their apparel production lines (Brady et al 2000).

11.1 Mass Customization's Pitfalls

Despite great potentials of mass customization to serve both process efficiency and diversified customer requirements, it is not easy to be implemented. The Yankee Group, specializing in the analysis of trends in strategic planning, technology forecasting and market research, studied the mass customization adoptability and discovered that it is not economically feasible for most industries. Mass customization makes more sense for such items as cars and computers or products where customization appeals to personal tastes, health concerns, or utilitarian advantage.

Returns of customized products can be problematic. Theoretically, customers who have specifically chosen products they really want will be less likely to return them, but some may want to. Once the retailers have products returns, they will find it very difficult to release such special products. This problem may be resolved by prohibiting the returns of customized goods for any reason other than defects.

The biggest impediment to mass customization is that most company supply chains cannot handle it. Supplier systems are often designed and optimized for producing predetermined amounts of stock rather than responding to direct demand. Many do not even incorporate modern supply-chain management practices. The solution to these supply-chain problems is a compromise between mass production and mass customization, wherein companies initially manufacture general products and then configure them to meet specific customers demands later in the process (Mello 2001).

12. IMPACTS ON NPD STRATEGIES

Joseph Pine (1993) has described a paradigm shift from mass production to mass customization as shown in figure 2-5.

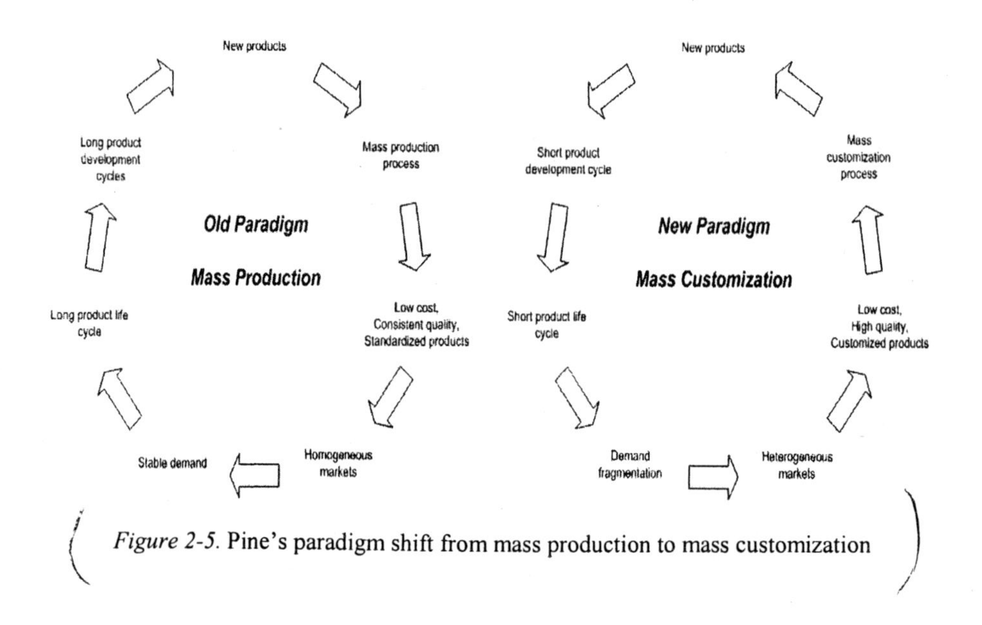

Figure 2-5. Pine's paradigm shift from mass production to mass customization

Pine used market turbulence - characterized by unstable and unpredictable demand levels; heterogeneous desires; price, quality and style consciousness; high levels of buyer power; competitive intensity; product differentiation; and market saturation, to identify the forces shifting the focus of manufacturing from mass production to the new paradigm of mass customization. A number of companies in the automotive, electronics, and the apparel industries have successfully made the transition of at least some of their business to a true state of mass customization. Existing companies evolve into a mass customization system in response to market turbulence.

Looping from the creative/inventive stage through mass production to continuous improvement and finally arriving at a true state of mass customization (Pine 1993).

To validate this assumption, a hypothesis is derived and to be empirically verified in chapter 5.

Proposition 3: *The market conditions characterized by short product life cycle and unstable demands have influence on NPD strategies.*

The independent variables used in this study were [1] percent of product life cycle reduction during the past decade, %PLR; and [2] percent more demand fluctuation in last 10 years, %MDF.

Within the dependent variable camp, three factors were selected to study. First was percent of innovative products a company introduced into its market. Among five types of product development strategy, only new-to-the-company and new-to-the-world products are classified as innovative products. As a result, the first variable was the summation of values in new-to-the-company and new-to-the-world categories. Second dependent variable was the direction of marketing strategy whether to move toward price competition or product differentiation. The third dependent variable was the percent product development time reduction (%PDTR).

Based on to two driving factors and three driven outcomes, the proposition 3 can be expanded to 6 postulates as listed by follows:

Proposition 3a: Ratio of innovative products launched by a company positively relates to product life cycle reduction.

Proposition 3b: Companies base marketing strategy more on product differentiation rather than price competition when product life cycles shrink.

Proposition 3c: Product development times change in conjunction with product life cycles.

Proposition 3d: Ratio of innovative products launched by a company positively relates to product demand fluctuation.

Proposition 3e: Companies base marketing strategy more on product differentiation rather than price competition when product demands are more fluctuating.

Proposition 3f: Product development times are reduced when product demands are more fluctuating.

All of these hypotheses were validated with the survey data. The results are discussed in details in the chapter 5.

13. TOTAL COST OF OWNERSHIP IN A SUPPLY CHAIN CONTEXT

Where many companies once had monolithic vertical supply structures--owning facilities that produced parts and subassemblies, most now focus on their core business, often just developing and marketing their end product. Parts and subassemblies are manufactured, and often designed, by suppliers and vendors. Some do not even view themselves as manufacturers anymore but as service providers, providing a linkage between end consumers and the manufacturer. This is particularly true in the electronics and automotive industries, where the modularity of their products allows for the easy outsourcing of manufacturing (Tully 1994), though recent downturns in the electronics market are pushing contract manufacturers to diversify into other industries (Serant 2002).

This has led to the development of supply chains, interconnected and highly dependent networks of companies that take products and services from concepts and raw materials all the way to the end customer. These organizations really came to prominence in the 1980s, as companies sought to enhance competitiveness by containing costs, enhancing product value, compressing the time to market, creating channel efficiency and becoming more responsive to customers (Cavinato 1991). Figures 2-6 and 2-7 show a generic product supply chain and an example of books supply chain.

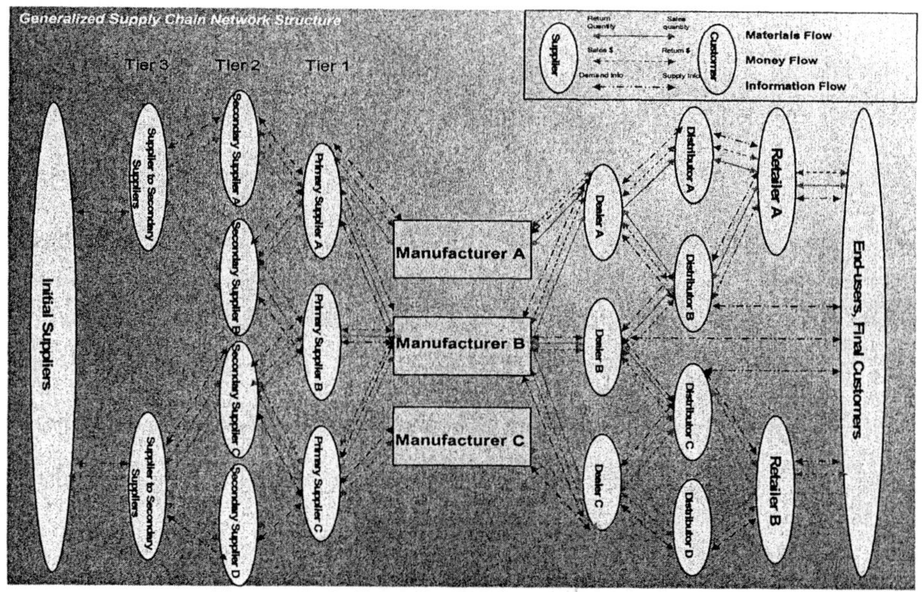

Figure 2-6. Generic Product Supply Chain

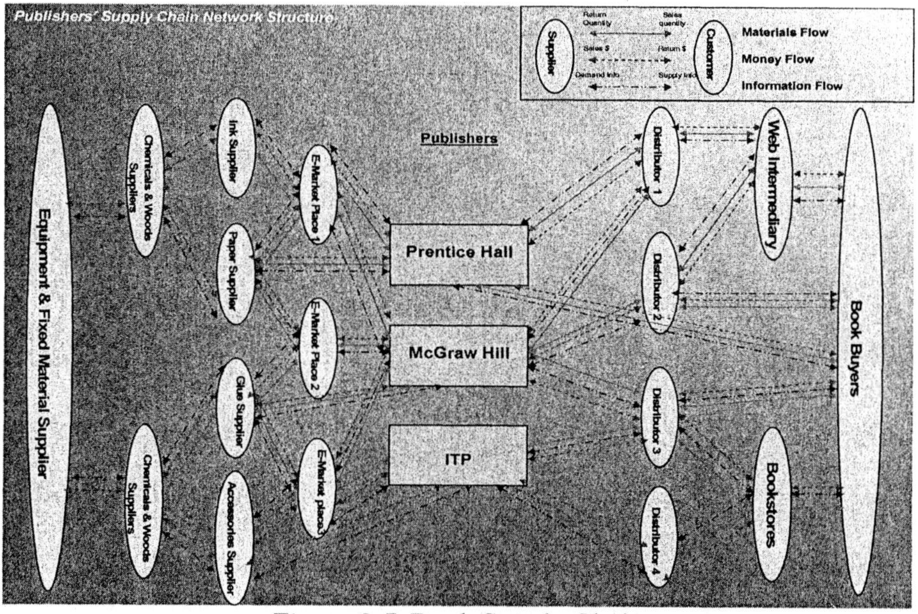

Figure 2-7. Book Supply Chain

As the outsourcing trend continues, companies won't merely compete against each other, entire supply chains will. The new strategy of supply chain will use a new set of principles: the only entity that puts money into the supply chain is the end customer, and the only viable solutions ensure that every element of the supply chain profits. Therefore, supply chain management is about economic value and total content; price is not the only issue anymore (Handfield 2002).

In order to stay competitive, the companies that sit within these supply chains will need to understand and influence the costs within their supply chains. Marien & Keebler (2002) have suggested that there are six stages of cost focus in a company's supply chain.

- Stage 1: Functional Cost Minimization—functional areas look to reduce their individual costs, often with cost penalties elsewhere in the system.
- Stage 2:Lowest Delivered Cost—company looks to minimize costs on acquired and delivered goods and services, often by looking at trade-offs in purchasing, transportation and asset management.
- Stage 3: Total Cost of Ownership—company begins to examine inventory and asset carrying costs.
- Stage 4: Enterprise Value-add Costs of Sales—company begins to look at the costs beyond the mere costs of material ownership, e.g. sales & marketing, engineering, technical support, IT, etc.
- Stage 5: Interenterprise Value-add Cost with Immediately Adjacent Trading Partners—examines trade-offs and best working relationships with the immediate customer and supplier in the supply chain.
- Stage 6: Lowest End-user Delivered Supply-chain Cost—examines trade-offs and services between all members of the supply chain with focus on the ultimate end user. (Marien & Keebler 2002)

The supply chain cost perspective is migrating towards a total view of the system both upstream and downstream, beyond just purchasing and inventory costs (Handfield 2002).

In this environment, purchasing takes on a critical role. In most supply chain situations, procurement organizations become the manager of the wide and varied relationships with vendors, channeling communications from customer to vendor and leveraging buying power to the company's advantage (Cavinato 1991). They manage a greater and greater amount of the company's spend. Cavinato (1992) estimated that the cost of procurement specification and acquisition is 55-60% of the total cost in manufacturing firms. Carr & Ittner (1992) estimate that purchase materials, components and subassemblies represent over 70% of manufacturing expense; Ellram (1995) placed it at 63.5% of total cost in manufacturing firms and 25% in non-manufacturers . More recently, Handfield (2002) puts the cost managing the supply chain at 56% of revenue at average manufacturing companies and increasing as one looks at more high technology industries. Also, each dollar cut from the cost of purchasing generates the same bottom-line effect as increasing sales by $17 (Handfield 2002). In a competitive worldwide market with thinning margins, purchasing finds itself under pressure to reduce procurement costs.

In highly competitive worldwide industries, companies need a strategic edge to gain and maintain market share. Traditionally, companies choose to compete using one of five factors: cost, dependability, innovation, quality, and flexibility. Effective control of bottom-line cost can provide opportunities to exploit a number of these factors. Obviously, lower costs can allow a company to offer lower prices without sacrificing profit margins. Alternately, lower costs means that more profits can be invested on developing new innovative products or on improving product quality and dependability.

Because procured components and services make up such a large part of the overall cost of a product in these competitive industries, it makes sense to examine them and try to control their contribution to the overall cost. Total cost of ownership (TCO) is a process of analyzing supply chain activities and their associated costs. Ellram & Siferd (1993) proposed TCO, but the general concept has been around prior to 1993 under a number of different names: total cost (Cavinato 1991; Cavinato 1992), life cycle costing (Jackson & Ostrom 1980), cost-based supplier performance evaluation system (Monczka & Trecha 1988), cost of ownership (Carr & Ittner 1992), zero base pricing (Burt, Norquist & Anklesaria 1990), and product life cycle costs (Shields & Young 1991). All of these concepts are structured around three basic supporting ideas: that cost must be examined from a long-term prospective beyond just the initial price, that purchasing must consider the effects of other business functions on the value of a specific purchase, and that purchasing must understand the cost impacts of all purchasing activities (Ferrin & Plank 2002). For example, Ellram (1994) estimated that the purchase price only accounts for 35% of the TCO in manufacturing equipment. This concept is not new to the 1980s and 1990s, either; purchasing management sources from as far back as 1928 have been emphasizing the importance of looking beyond initial purchase prices (Ellram & Siferd 1993). The US Department of Defense (DOD), in particular, started using total cost principles for its procurement activities starting in the early 1960s (Shields & Young 1991). Prior to the outsourcing trend that exists today, much of this early focus on life cycle costing was aimed at equipment and capital purchases, usually examining maintenance and energy costs for two or more alternate equipment purchases (Jackson & Ostrom 1980).

Zero base pricing (ZBP) is a total cost method developed and trademarked by Polaroid in the early 1980s. It is based on the all-in-cost concept where the "all-in-cost" equals the acquisition price plus the "all in-house" costs. The all in-house costs are described as all of the costs needed to convert the purchased material to the finished product, including any costs from the field failure of the final product due to defects in the original purchased materials. In zero base pricing, each cost area consists of avoidable and unavoidable cost components. ZPB works off of identifying the avoidable cost components and eliminating them. The model looks at the supplier cost areas of profit, general & administrative costs, factory overhead, labor, and materials and at the in-house cost areas of customer returns & lost sales, warranty, service & field failures, scrap, process yield losses, rework, lost production, production, storage, inspection & testing, and incoming transportation. Avoidable cost reduction is carried out by looking at six cost and quality drivers: tolerances, specifications, materials, the process and the requirements placed on it, design for automation, and ease of manufacture. The system places a heavy emphasis on supplier negotiation and early supplier involvement in the design process to maximize the cost savings (Burt, Norquist & Anklesaria 1990).

1980s cost management literature defined life cycle costing (LCC) as all the costs that the producer will incur over the product's life cycle; Shields & Young (1991) expanded this definition also include all of the costs incurred by the end consumer and called it product life cycle costing (PLCC). Though they didn't go into great detail defining the specific mechanics of the system, Shields & Young pushed the idea from mere cost estimating tool to cost management system. They are describing a product life cycle cost management (PLCCM) system which takes a product-centric view of costing, focused on creating long-term success in competitive markets through the production of innovative, quality products with short lead and delivery times. As such, the system they describe has a very strategic focus, attempting to align the company structure and employee behaviors to the PLCCM in order to maximize its benefit (Shields & Young, 1991).

Cavinato (1991, 1992), whose focus appears to be on supply chains and how the different agents in the supply chain add value to the final product, develops a model called total cost. Like TCO, total cost looks at all of the costs associated with a product, but it does so from the perspective of a make/buy-type of decision—i.e. comparing the costs at the supplier to the costs at the customer to determine which can make the product for the lowest total cost. It does this by comparing six cost factors: lowest labor rate, most effective process, most capital available, lowest cost of capital, highest tax rate, and most investment tax credits and depreciation available for use (Cavinato 1991). He later creates a total cost/value model which advocates collecting costs in ten key areas: traditional basic input costs, direct transactional costs, supply relational costs, landed costs, quality costs/factors, operations/logistics cost, indirect financial costs, tactical input factors, intermediate customer factors, and strategic business factors, but he doesn't discuss any mechanism for actually estimating those costs.

Carr & Ittner (1992) codified a cost model, called cost of ownership, that included five main cost items: purchase price, costs of purchasing, costs of holding, costs of poor quality, and costs of delivery failure. Like Cavinato, they do not recommend a specific costing mechanism, but instead give examples of four manufacturers' cost of ownership systems. Each example involves a cost-ratio method where the total of the five cost areas is divided by the purchase price to get a cost index for each supplier. The user can then multiple this index against future purchase prices to estimate the probably cost of ownership (Carr & Ittner 1992). Monczka & Trecha (1988) had earlier proposed a similar cost-ratio system that they called cost-based supplier performance evaluation system (CBSPES). In CBSPES, the purchased price is added to "non performance costs", mostly quality and logistics costs, and divided by the purchase price to get an index which can be averaged to achieve an overall one for each supplier. CBSPES advocated the use of events with standardized costs to determine non performance costs. For example, scrapping of lot of material or returning materials to a supplier has a standard cost to the quality department. CBSPES is unique for a total cost system in that it utilized a second rating system to capture subjective service factors that might not have an easily identifiable cost such as willingness to share data and responsiveness of communications (Monczka & Trecha 1988).

Originally, Ellram and Siferd (1993) describe TCO as "all costs associated with the acquisition, use, and maintenance of an item". To aid in determining these activities, they break them into six broad categories: quality, management, delivery, service, communications, and price. They differentiated TCO from zero base pricing by focusing TCO on internal customer costs while zero base pricing is focused on costs at the supplier (Ellram & Siferd 1993). It differs from Cavinato's total cost concept in that it actually provides a more specific method for collecting the costs using activity analysis. Ellram (1993) soon revised TCO to include capital equipment, maintenance, repair, and operating supply (MRO) items, and services in addition to purchased components and materials. She also recommended organizing activities into generalized pretransactional, transactional, and posttransactional categories, getting away from the previously described six functional categories. Another change was the suggestion that only significant cost components, those that account for most of the probable TCO, warrant tracking (Ellram 1993).

More recent articles regarding TCO deal not so much with the theory behind it and its importance, but with the disappointing implementation of it. Recent studies (Milligan 1999; Ellram & Siferd 1998; Ferrin & Plank 2002) have indicated that while many companies utilize TCO principles, they currently don't use them systematically or evenly. Often, TCO usage is an informal process where data has to be collected from numerous sources, often manually or with less-than-optimized IT systems (Milligan 1999). This informal process often produces vague, untrustworthy and inaccurate results which inspire less confidence in the TCO system (Milligan 1999). Many companies don't apply it to all of their purchasing decisions; some apply it routinely while others used it only for high priority or high cost items (Milligan 1999; Ellram & Siferd 1998). Sometimes, rather than using a general TCO model, companies develop unique TCO models for specific purchasing decisions (Ellram 1994). There's a great deal of evidence that a single TCO model with specific set of cost drivers cannot be applied to all TCO users, but instead, TCO guidelines should be general because they will be customized by the user (Ferrin & Plank 2002; Ellram & Siferd 1998).

TCO focuses on analyzing supply chain activities and their associated costs. This requires the use of systematic tools to determine activities, estimate their costs, and manage them. Over time, numerous cost estimation methodologies have developed. Zhang & Fuh (1998) describe six different costing methods:

- Traditional detailed-breakdown cost estimation—a detailed summation of all the costs (material, direct labor, overhead, etc.) that occur in the manufacturing process.
- Simplified breakdown cost estimation—simplified cost method for use prior to the development of the detailed process plan. It assumes use of optimum manufacturing method regardless of the actual equipment and process that will actually be used. Because of its use of empirical equations, it's often difficult to correlate with actual costs.
- Group technology (GT)-based cost estimation—utilizes a base cost for a group of similar products and a set of cost variables such as size, length,

or number of features and then establishes linear relationships between the final cost and the variable cost factors.

- Cost estimation based on cost functions or cost increase functions—useful for a similar group of products, this method combines the technical and economic elements into a single functional equation of the overall cost. Coefficients for this equation are then determined by feeding historical data into a regression analysis.
- Activity-based cost (ABC) estimation—described in detail below
- Neural network approach to early cost estimation—Uses a back-propagating neural network (a series of layered algorithms with an input layer, an output layer, and a number of "hidden" layers between them equal to the number of inputs) which can be set up with historical cost data and then adjusted or "trained" as additional cost information becomes available or the manufacturing situation changes.

Historically, use of the traditional costing method prevailed; cost estimates used a cost allocation system where overhead costs were applied to the manufacturing process on the basis of direct labor hours or direct machine hours. That cost was then added to the direct labor cost and the raw material cost to arrive at a total cost for the product. This produced adequate cost estimates when direct manual labor made up the major portion of the cost. In recent years, however, more production has become automated, reducing the amount of direct manual labor on a product. In the 1990s, conventional wisdom held that direct labor comprised only 5-10% of a typical products cost and over 50% of the cost could be attributed to overhead (Porter 1993). The continued use of traditional cost systems with a direct labor allocation basis now distorts cost estimates in many situations.

Activity based costing (ABC) is an activity-focused cost estimation and analysis method which seeks to provide a more accurate allocation of indirect overhead costs and complements the activity focus of TCO well (Porter, 1993; Ellram 1995). This ABC methodology can be used to analyze the supply chain related costs both at the purchasing company and at the supplier. In fact, as companies push to outsource more and more of their components and services, it becomes more important that they apply ABC analysis to their suppliers' activities and understand their costs.

The development of ABC takes place in the context of the Harvard Business School in the early and mid 1980s where American business was under threat from global markets, primarily the Japanese (Jones & Dugdale 2002). Independent studies by Robin Cooper, Robert S. Kaplan, and H. Thomas Johnson and the Computer-Aided Manufacturing, International

(CAM-I) organization all followed along similar threads, looking at revamping cost accounting systems to better account for overhead costs (Jones & Dugdale 2002). In a series of articles, Kaplan argues that management accounting methods and practices had become irrelevant and obsolete in modern manufacturing. This work cumulated in a 1987 book with Johnson, *Relevance Lost*, which details the decline in management accounting (Lukka & Granlund 2002). At the same time articles by Kaplan and Cooper argued that traditional costing systems were also ineffectual, producing bias information in a modern manufacturing environment. These paved the way for ABC which Kaplan and Cooper introduced in a *Harvard Business Review* article in April 1988, based largely on their work with companies like John Deere, Scovill, Hewlett-Packard, and Siemans where ABC-type systems were being developed (Lukka & Granlund 2002). The original definition ABC, sometimes called "first wave ABC", assumes that activities are controlled, allocation of cost is exact and complete and the results are described as "more accurate" (Jones & Dugdale 2002). A series of articles by Cooper, outlined the ABC process in detail (Cooper 1988a, 1988b, 1989a, 1989b).

At its central core, the concept of ABC revolves around developing a more accurate cost model of manufactured goods. A traditional volume-based cost system can generate distortions and bias when applied to diverse production volumes, part sizes, complexities, materials, and set up requirements. Use of ABC is aimed at eliminating these biases. While traditional costing systems focused on the product and usually allocated costs on the basis of direct labor hours, machine hours, and material dollars, ABC is focused on activities and how products use activity resources. It generates more accurate results by allocating costs from a wide variety of cost bases (Cooper 1988a).

The following chapter discusses research methodology and tools used in the study.

Chapter 3

RESEARCH METHODOLOGY

The methodology of research used in this study basically includes literature survey, industrial survey, and verification by statistical techniques. Nonetheless, there were also some limitations incurring from the industrial survey results that may lead to misinterpretation discussed at the end of the chapter.

1. LITERATURE SURVEY

The first step in developing a body of knowledge essentially begins with searching previous research to understand how far the people in the field of interest have gone through the issue. The CLIC Libraries (the book indexing system) and the Internet access systems at the University library were utilized to search for books, articles, thesis, and researches. Since this research focused on new product development, the keywords used in searching for literature were among: *New product development; New product; Product development; Innovation;* and *Innovation management.* In addition, other related terms to the NPD process, for instance, *Strategy; Competitive advantage; Time to market; Knowledge base; Organization learning; Product life cycle management; Mass customization;* and so forth were also used as the keywords in searching.

2. INDUSTRIAL SURVEY

The industrial survey and data analysis were probably the most sophisticated and time taking processes in this research project. Once the NPD topic was selected for study, there was not a particular theme for the research yet. The development of industrial survey (Appendix B) reflected this approach as the included questions covered a wide range of topics on industry practices and approaches. Due to a large number of questions (174 questions in total), the survey material was iteratively designed in a form of questionnaire containing mostly multiple-choice and ranking questions and a few questions in the form of short-answer in order to speed up the survey completion time. The survey was expected to explore the strategy, tactics, and practices that the selective manufacturers in various industries are implementing in the present day. The survey instrument was comprehensive and covered a broad range of aspects having potential affecting NPD strategies and practices. It has questions constituting 12 categories as shown in Table 3-1 (figures in brackets indicate number of questions on that particular topic).

Table 3-1. Multidisciplinary aspects in the NPD questionnaire

Topic	Objectives / Contents
• *General background information* (17):	Indicating size of company in terms of revenue and number of employees, industry type, level of innovation and competition, product life cycle, and new product idea surviving percentage[2].
• *Product development strategy* (6):	Describing objectives and positions of new products introduced to markets.
• *Product success* (5):	Showing the success of new products launched into market in terms of revenue, market share, profit margin, customer satisfaction, and innovativeness.
• *Customer interface issues* (21):	Studying how much customers are involving in NPD process in terms of customer requirement and feedback evaluation, customer's accessibility to design data and project team, tools used to specify customer needs, and characteristics of market-competition landscape.
• *New product development timeliness and schedules* (20):	Focusing on how well companies manage their NPD projects by means of project timeliness including tools used for measuring project success and performance.
• *New product development processes* (23):	Revealing how formally companies handle NPD process through well-established procedures like stage-gate, modular technique, FMEA, risk mitigation planning, and collaborative engineering tools.
• *New product research* (8):	Showing how much research activities are carried out in NPD process and importance of research contributing to company's innovations.
• *Teamwork and leadership* (10):	Describing how NPD project team leaders and members are formed as well as rewarding and performance evaluating methods.
• *Human resource development* (6):	Explaining how well companies pay attention on human resource development especially for NPD projects.
• *Technology deployment* (5):	Telling whether companies have a formal process to deploy and manage NPD-relating technologies.
• *Cost / Profit margin / Return* (15):	Studying how effective companies monitor and control project and product costs, returns on investment, cash flow, and capital expenditures.
• *Market turbulence* (25):	Exploring effects of market-competition conditions and customer behaviors changing over time on NPD strategy formulation and management.

[2] Number of original ideas decreases as they reach each stage of new product development processes, from idea generation to commercialization, because some ideas cannot prove feasible at those stages. The surviving rate shows the ratio of remaining ideas out of total 100% ideas originally generated.

Survey distribution and getting responses were challenging issues faced in this research study. The survey was targeted to companies manufacturing tangible products in various industries nation wide. Top ten U.S. companies in terms of revenue size in each standard industrial code (SIC): - whose product categories are such tangible products as automobile, computer components, household appliances, electronics consumer goods, semiconductor, tools and machinery, and medical devices; were invited to participate in the survey. 45 respondents from 31 companies in automobile; biotechnology and drugs; computer storage; construction; defense; electronics and electrical components; foods; industrial machinery; instrument; medical devices; and printing and publishing industries returned completed survey instrument. The survey response ratio was 22.1% of all survey questionnaires mailed to firms.

The illustration of surveyed companies' demography classified by industry types is shown in Figure 3-1. The sizes of participating companies in terms of number of employees varied by 29%, 42%, and 29% for small sized (< 1,000), medium-sized (1,000-10,000), and large-sized (>10,000) respectively (Figure 3-2). When considering revenue distribution of surveyed firms, it was surprising to find that most surveyed firms (55.6%) had high annual revenues (>$1,000 million); medium-level firms (with $100 - $1,000 million in revenues) were accounting for 24.4%; and low-income companies (with less than $100 million in revenue) were contributing to only 20%. Figure 3-3 shows this revenue distribution. Product types of surveyed companies can be categorized into industrial products 56% and consumer products 44%.

One may argue that the survey results might have been biased due to the geographic distribution of participating companies: - all located in U.S. Midwest area. However, the values and practices a company possesses are mainly influenced by its headquarter. Therefore, despite the responses were from the same region, the survey results can represent values and practice of local, regional, national, and international companies. The respondents ranged from engineer-level (process engineers, product development engineers, design engineers, etc.) through manager-level (quality managers, production managers, project managers, etc.) to director-level (R&D director and vice president of operations). These diversities in persons, companies, and industries were anticipated to reveal common practices across the broad range of industries.

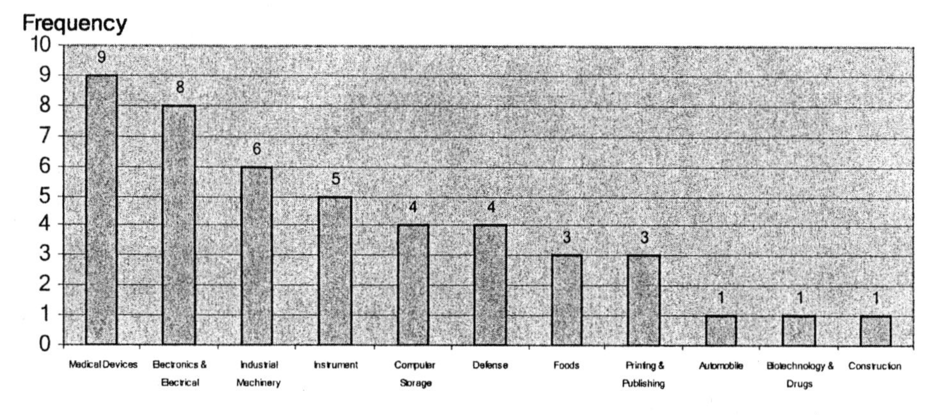

Figure 3-1. Distribution of survey companies by industries (Some companies may have more than one respondent)

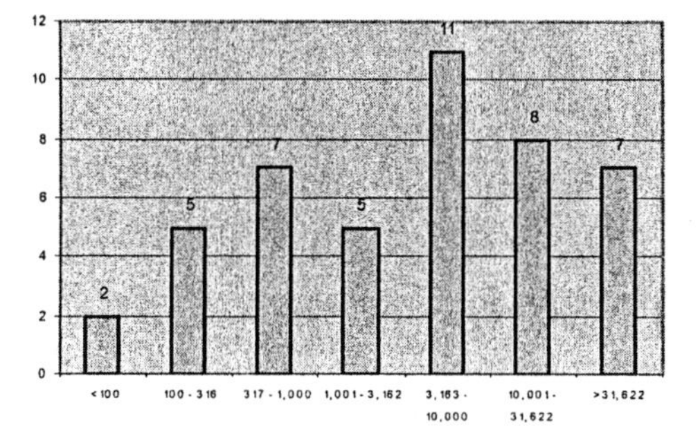

Figure 3-2. Demography of survey companies by number of employees

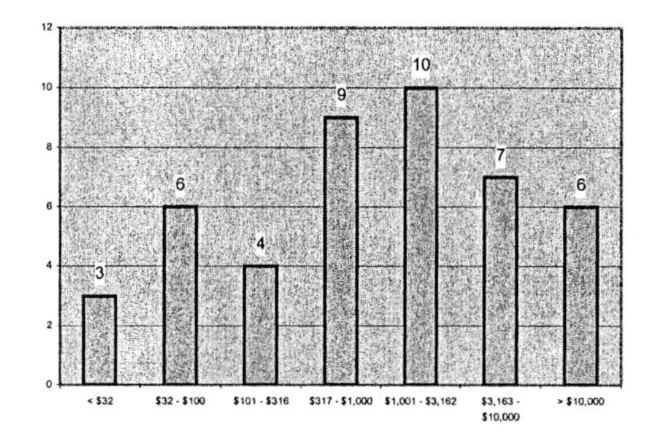

Figure 3-3. Distribution of survey companies by revenues

3. SURVEY DATA ANALYSIS

As the questionnaire had been designed to facilitate respondents to answer quickly, the data retrieved from the responses therefore were mostly in numeric and yes/no types, which are easy to be analyzed in spreadsheet and statistical analysis programs. Excel spreadsheet and SPSS 11.0 statistical analyzer were mainly used as the major software to analyze the data and depict results in graphs and charts. Excel was used to arrange data into tables; calculate sum, average, and other functions; and create charts. In forms of column-bar charts, line charts, pie charts, X-Y scattering charts, box plots, and combination, graphical illustrations were mostly used to portray data from numeric to pictures that help viewers see distribution and trends easily. SPSS 11.0's functions utilized in this research study included cross tabulation, analysis of variance (ANOVA), correlation analysis, and regression analysis.

4. RESEARCH STUDY PROPOSITIONS AND VERIFICATION

After the existing body of knowledge relating to NPD from books and journals plus initial survey data analysis had been reviewed, new hypotheses were proposed to generate new concepts adding into the knowledge obtained from previous studies. The study could be summarized in three main themes: [1] typology of NPD strategies; [2] roles of organization learning and knowledge management in NPD; and [3] market turbulence effects on NPD strategies.

Variables used in the theme 1:

To verify the proposed hypothesis, both descriptive and statistical techniques were used to seek for correlations between NPD strategies and product success. Based on the questions in Product Success section, the product success was defined as the respondents' perception on how successful their new products can contribute to revenue growth, profit margin, market share, customer satisfaction, and innovative features. As a result, the summation of scores ranked for the question in this section was used as the dependent variable of this study. For the driving factors, the answers to questions 1 and 3 in the New Product Strategy section were selected to stand for customer-responsive strategy and innovative strategy accordingly. The summation of scores of both questions served as the third independent variable that represented the combined strategies.

Variables used in the theme 2:

The NPD process-related success factor in the questionnaire is the response to question 24 of Market Turbulence section that compares NPD cycle times today with that of 10 years ago. The ratio of the cycle time difference over the cycle time of last 10 years determines percent of NPD cycle time reduction. The ratio was used as the process-related success factor in this study.

The project on time ranking, corresponding to question 6 of NPD Timeliness and Schedules that asked the respondents to rank ability of their companies on completing NPD projects on time, was chosen to represent the project-related success.

For the independent variable, the summation of results from questions relating to organization learning and knowledge management was selected for the study. The designated questions included;

- Question 2 of NPD Processes section: companies offer training for standard NPD procedures.
- Question 12 of NPD Processes section: continuous learning is incorporated into project teams.
- Question 1 of Human Resource Development section: training needs assessment.
- Question 2 of Human Resource Development section: training gap analysis.
- Question 3 of Human Resource Development section: monitoring training needs and status.
- Question 5 of Human Resource Development section: minimum set of training for new employees.
- Question 6 of Human Resource Development section: companies pay for trainings and tuitions.

Variables used in the theme 3:

The independent variables used in this study were: [1] percent of product life cycle reduction during the past decade, %PLR; and [2] percent more demand fluctuation in last 10 years, %MDF. The %PLR was computed from the responses of question 15 in the Market Turbulence section according to following equation:

% Product life cycle reduction= (Product life 10 years ago – Product life today) / Product life 10 years ago * 100%

The %MDF was calculated from the answers of questions 1 in the same section as %PLR. The formula of obtaining the %MDF was also similar to %PLR as follow.

% More demand fluctuation = (Demand fluctuation 10 years ago – Demand fluctuation today) / Demand fluctuation 10 years ago * 100%

Within the dependent variable camp, three factors were selected for the study. First was percent of innovative products a company introduced into its market. This figure can be obtained from the responses of question 6 in the Product Development Strategy section. Among five types of product development strategy, only new-to-the-company and new-to-the-world products are classified as innovative products. As a result, the first variable was the summation of values in new-to-the-company and new-to-the-world categories.

Second dependent variable was the direction of marketing strategy whether to move toward price competition or product differentiation. The responses of question 12 in the Market Turbulence section indicated this variable following this equation.

% More product differentiation = (Marketing strategy 10 years ago – Marketing strategy today) / Marketing strategy 10 years ago * 100%

Pine (1993) described that a consequence of new product development paradigm, characterized by short product life cycle, is shorter product development time. This study took this statement to verify empirically with the surveyed data. Thus, the third dependent variable was the percent product development time reduction (%PDTR) derived from the answers of question 24 in the Market

Turbulence section. Following is the calculating formula of %PDTR.

% PDTR = (NPD cycle time 10 years ago – NPD cycle time today) / NPD cycle time 10 years ago * 100%

In addition to statistical analyses, the literature and previous researches were also repeatedly reviewed in order to advocate or compare and contrast the survey results and findings.

The next chapter discusses the survey results and analyses.

Chapter 4

RESULTS AND ANALYSES

This chapter describes the outcomes of the survey in general terms. The description will proceed along twelve sections of the survey material. Besides normal descriptive and statistical analyses of survey data, some topics of interest to readers will be emphasized supported by related research articles in order to expand the knowledge of the subject matter.

1. GENERAL BACKGROUND INFORMATION

According to Table 4-1 and graphics of distribution patterns of surveyed companies in Figure 4-1 and Figure 4-2, one might notice that most consumer-product manufacturers in this survey samples were superior to the industrial-product manufacturers in terms of both employment and incomes. To be precise, nine consumer-product makers had employees over 10,000 head counts while only six of all counterparts had the same manpower. In terms of income, 15 consumer-product manufacturers had revenues more than $1 billion whereas only 8 industrial-product firms earned at the same level.

Table 4-1. Descriptive statistics of surveyed companies

Attributes	Categories	Consumer Product		Industrial Product		Total	
		Count	%	Count	%	Count	%
Ownership	Public	17	85%	14	56%	31	69%
	Private	3	15%	11	44%	14	31%
		20	100%	25	100%	45	100%
# Employees	< 100	0	0%	2	8%	2	4%
	100 – 316	0	0%	5	20%	5	11%
	317 - 1,000	3	15%	4	16%	7	16%
	1,001 - 3,162	1	5%	4	16%	5	11%
	3,163 - 10,000	7	35%	4	16%	11	24%
	10,001 - 31,622	7	35%	1	4%	8	18%
	> 31,622	2	10%	5	20%	7	16%
		20	100%	25	100%	45	100%
Revenue (million)	< $32	0	0%	3	12%	3	7%
	$32 - $100	2	10%	4	16%	6	13%
	$101 - $316	1	5%	3	12%	4	9%
	$317 - $1,000	2	10%	7	28%	9	20%
	$1,001 - $3,162	8	40%	2	8%	10	22%
	$3,163 - $10,000	3	15%	4	16%	7	16%
	> $10,000	4	20%	2	8%	6	13%
		20	100%	25	100%	45	100%
Product Innovation	Fast changing	12	60%	10	40%	22	49%
	Slow changing	5	25%	10	40%	15	33%
	N/A	3	15%	5	20%	8	18%
		20	100%	25	100%	45	100%
Competition	High competition	15	75%	19	76%	34	76%
	Low competition	0	0%	0	0%	0	0%
	No competition	1	5%	1	4%	2	4%
	Many M&A	2	10%	1	4%	3	7%
	N/A	2	10%	4	16%	6	13%
		20	100%	25	100%	45	100%

Continue

Attributes	Categories	Consumer Product		Industrial Product		Total	
		Count	%	Count	%	Count	%
# Products	1 Product	0	0%	1	4%	1	2%
	10 Products	3	15%	6	24%	9	20%
	100 Products	7	35%	7	28%	14	31%
	Legacy products	7	35%	6	24%	13	29%
	N/A	3	15%	5	20%	8	18%
		20	100%	25	100%	45	100%
Innovation Types	Incremental	12	60%	19	76%	31	69%
	Radical	7	35%	4	16%	11	24%
	Both	1	5%	2	8%	3	7%
		20	100%	25	100%	45	100%

More than 80% of survey responses revealed that their companies were operating in highly competing and consolidating environments. Only two companies claimed their unique, legacy products facing no competition in their industries. Only half of all respondents thought that their industries offered highly innovative products to customers. When considering pathway of technology and innovation changes, most responses (69%) used incremental methods to develop new products by extending from existing technologies. Only a quarter of all respondents claimed their companies usually developed new products through platform changes or radical technologies. While famous strategists from Massachusetts Institute of Technology (MIT) and Society of Organization Learning (SoL) like Peter Drucker, Peter Senge, and Gary Hamel all encourage business leaders to focus on radical or discontinuous innovations for sustained competitive advantage, it seems too difficult for companies to pursue as it is shown in the results at the bottom of Table 4-1.

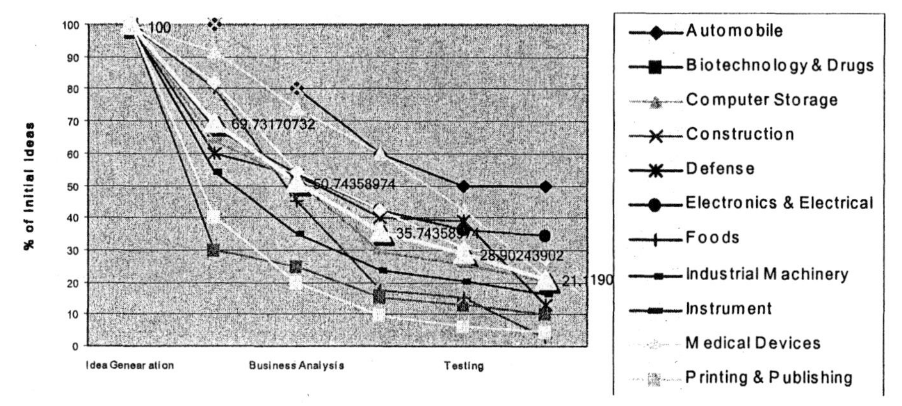

Figure 4-1. New product idea mortality curves of surveyed companies by industry

New product idea mortality curve has been utilized since the first NPD research in 1968 by Booz, Allen, and Hamilton (BAH) and has continually been used to compare the new products idea screening over time. The mortality curve represents the progressive rejection of ideas or projects through stages of the new product development process: - from idea generation to idea screening, business analysis, product development, testing and trial, and commercialization (Griffin 1997). Figure 4-1 presents idea mortality curves based on data from survey results by industry. The thick line is the average curve of all responses.

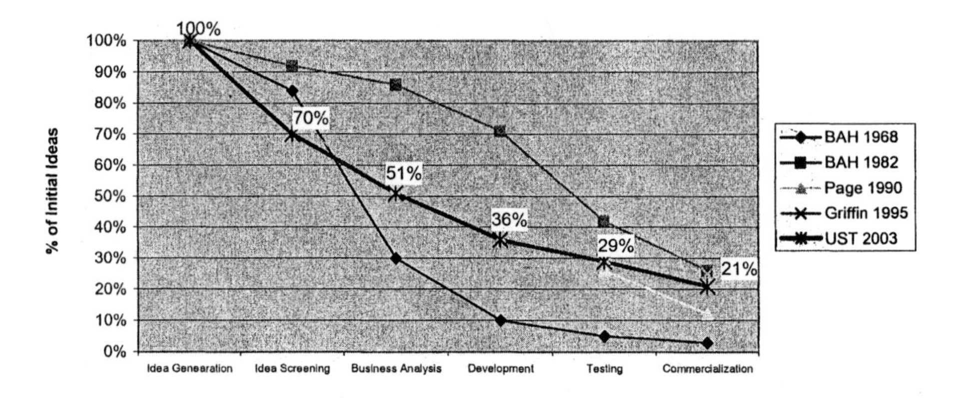

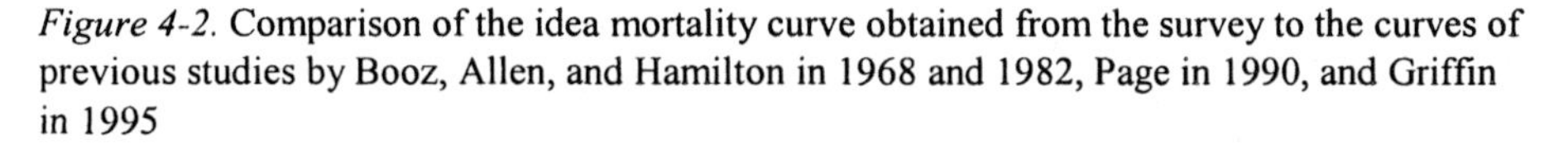

Figure 4-2. Comparison of the idea mortality curve obtained from the survey to the curves of previous studies by Booz, Allen, and Hamilton in 1968 and 1982, Page in 1990, and Griffin in 1995

When the new idea mortality curve is compared to those of previous studies as seen in Figure 4-2, the behavior of NPD projects at the present time is very much similar to the projects in the past decade (Griffin's). The only difference between Griffin's and this study results is in the commercialization stage where Griffin's had 24% (approximately one out of 4.2 ideas being commercialized) while this study had only 21% of initial ideas (one out of 4.8 ideas) launched into markets. By considering percent changes of new idea mortality as shown in Table 4-2, the idea screening through NPD stages has shifted from heavy elimination at later stages, as seen at BAH 1968 and BAH 1982 studies, to earlier stages, noticed from Griffin's and this study's results. Generally, projects that are eliminated earlier in the NPD process waste money, time, and resources less than those that fail at later stages (Page 1993, Griffin 1997). This survey suggests that only half of all new ideas enter development stage where large NPD expenses first accrue. This is an obvious confirmation that today's NPD projects are wasting less money on unsuccessful projects than those in former days.

Table 4-2. Comparison of ratios of surviving ideas at NPD stages showing % changes of each forwarding stages

Research Set	Idea Generation	Idea Screening	Business Analysis	Development	Testing	Commercialization
BAH 1968	100%	84%	30%	10%	5%	3%
% Change		16%	64%	67%	50%	40%
BAH 1982	100%	92%	86%	71%	42%	26%
% Change		8%	7%	17%	41%	38%
Page 1990					26%	12%
% Change						54%
Griffin 1995	100%	75%	50%	37%	30%	25%
% Change		25%	33%	26%	19%	17%
UST 2003	100%	70%	51%	36%	29%	21%
% Change		30%	27%	29%	19%	28%

2. PRODUCT DEVELOPMENT STRATEGY

This section of the survey aims to study two subjects under company strategies. First is, what NPD strategies among: [1] most fit to customer needs, [2] lowest product cost, [3] most innovative features, [4] first-to-market, and [5] leverage revenue, profit, and market share; companies used to define their business-operating directions. Another one is, how the firms position their new products based on newness to the companies themselves and newness to the markets. Referred to newness as Figure 4-3 shows, the positions of products launched in the market can be classified as:

- *Cost reduction*: Improving existing products for merely improved costs without anything new.
- *Market repositioning*: Developing existing products of company that target to new markets.
- *Add to existing lines*: New products that supplement company's established product lines.
- *New-to-the-company*: New products that allow a company to enter an established market for the first time.

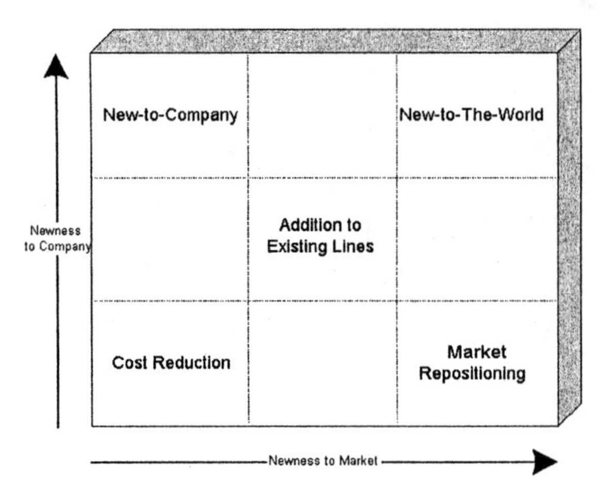

Figure 4-3. NPD strategy typology

- *New-to-the-world*: New products that create an entirely new market.

The framework implicitly positions the market strategy along the horizontal axis and technology strategy on the vertical axis since companies must depend on technology and innovations to create something new for them.

The survey respondents were asked to rank the objectives of new product development to their company's strategies with score from 0 (not relating at all) to 5 (fully relating). The responses to NPD strategy questions revealed that, on average, surveyed companies focused their NPD strategies to meet customer needs the most (85%), followed by leveraging profit and market share, being innovative, being first-to-market, and reducing product costs accordingly (Table 4-3).

Table 4-3. Summary of NPD strategies by industry

Industry	Most fit to customer needs	Lowest product cost	Most innovative features	First-to-market First-to-world	Leverage revenue/ profit/ market share
Automobile	80%	80%	100%	80%	100%
Biotechnology & Drugs	80%	20%	60%	60%	80%
Computer Storage	95%	80%	70%	80%	85%
Construction	100%	60%	80%	40%	80%
Defense	95%	45%	70%	27%	45%
Electronics & Electrical	75%	33%	55%	55%	63%
Foods	87%	60%	47%	60%	80%
Industrial Machinery	80%	64%	64%	52%	76%
Instrument	88%	28%	52%	72%	84%
Medical Devices	87%	53%	67%	44%	62%
Printing & Publishing	87%	73%	47%	40%	80%
Overall Average	85%	51%	61%	54%	71%

The second set of NPD strategy questions asked the respondents to allocate portions of new products developed in their companies into 5 categories of new product typology based on newness of companies and markets (figure 4-4).

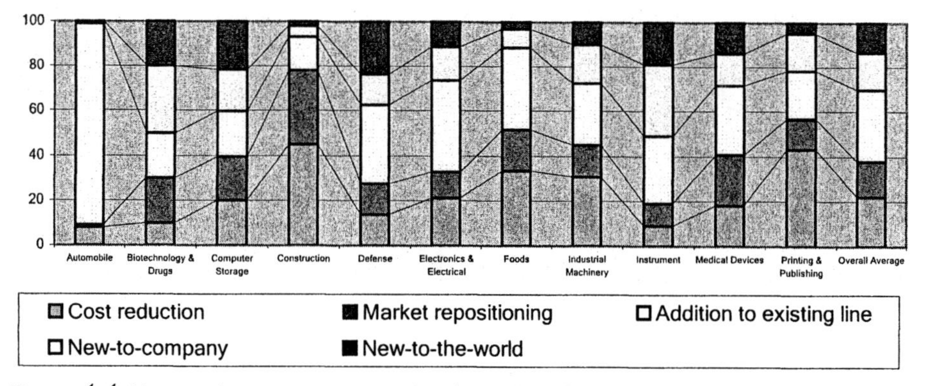

Figure 4-4. New product structure in each industry based on newness to the firms and markets

Figure 4-4 presents the results of survey on new product positioning to market in relation to innovation level of the launched products. On average, 31.68% of launched products fall into 'addition to existing lines' category. Surprisingly, cost reduction category, which has been defined as the least interesting issue in the previous survey set, is running up at the second place, with overall average of 22% for all products, followed by new-to-company, market repositioning, and new-to-the-world consecutively. This can be concluded from the results that more than half of new products from responding companies were not developed to be the first to company and/or market.

3. PRODUCT SUCCESS

This section was aimed to identify the results of product success in order to study their relationships with inputs of strategies and practices in other sections. The survey respondents were requested to rank their new product successes in terms of abilities of new products to meet profit targets, capture significant market share, generate significant revenue growth, provide unique benefits to customers, and bring the innovations to market. Again, the ranking scale ranged from 0 = never to 5 = always.

The results of each product-success category varied from industry to industry as shown in table 4-4. In aggregate, the success of new products in providing unique benefits to customers seemed to be outstanding. This result reflected the NPD strategy of surveyed companies focusing to best fit customer needs as discussed in the previous section. Overall results of other success categories were not far apart from the first one.

Table 4-4. Summary of new product successes ranked based on respondents' perception of their own products

Industry	Meet profitability	Capture market share	Generate revenue growth	Provide unique benefits	Be innovative to market
Automobile	100%	60%	60%	80%	60%
Biotechnology & Drugs	40%	40%	60%	60%	60%
Computer Storage	80%	85%	70%	65%	60%
Construction	60%	40%	40%	100%	80%
Defense	55%	60%	50%	90%	75%
Electronics & Electrical	48%	45%	43%	63%	50%
Foods	47%	33%	40%	33%	47%
Industrial Machinery	70%	60%	60%	73%	67%
Instrument	56%	64%	60%	80%	80%
Medical Devices	58%	60%	62%	62%	64%
Printing & Publishing	47%	47%	47%	67%	53%
Overall Average	58%	56%	55%	68%	63%

4. CUSTOMER INTERFACE ISSUES

This survey section explores the customer-involvement approaches that companies, looking for acquiring information from customer as inputs to their NPD processes, are supposed to exercise. Most responding firms have fairly implemented these tactics in a range of 50 - 70% of all time that they developed new products except some activities sensitive to proprietary information, such as customers have direct access to design database or customers directly involved in NPD process. Regarding the summary in table 4-5, individual practices are discussed as follows:

- *Customer requirement evaluation*: Companies in general performed the requirement evaluation 69% of all time. Notably, companies in defense industry fully exercised this practice at all time, while a company in automobile industry never reviewed end-user requirements at all.
- *Using product specification from customers*: On average, 58% of all time the surveyed companies used this practice with again 100% in defense and 0% in automobile industries.

- *Customer feedback reviews during NPD stages*: Surveyed corporation seemed to review customers' feedbacks at later stages rather than at the earlier ones as noticed from 58%, 58%, and 66% of involvements during specification, development, and testing and trial stages respectively.
- *Customers directly involved in NPD process*: Not surprisingly, only defense industry, which essentially has to work closely to heavy-weighted customers like Department of Defense (DOD) and military agencies, really involved customers in the NPD process that could bring overall score to 46% of all time.
- *Customers can access to detailed design information*: In addition to defense industry, computer storage makers also disclosed their design data to customers. However, overall practice was remaining low at 44% of all time.
- *Customers have access to prototypes*: Actually, this practice is generally required by most companies in order to collect feedbacks for improvement in a real production stage. As a result, the overall score rose to 60% of all time.
- *NPD project teams have direct customer contacts*: Companies in defense, biotechnology, and computer storage industries were leading in this practice since they worked closely with customers who were either end-users or other manufacturers in developing new products. The overall score was 64% of all time.

Table 4-5. Summary of responses to customer interface issues in NPD by industry

Industry	Automobile	Biotech & Drugs	Computer Storage	Construct	Defense	Electronics & Electrical	Foods	Industrial Machinery	Instrument	Medical Devices	Printing & Publishing	Overall Average
Customer requirement evaluation	0%	60%	65%	80%	100%	50%	80%	53%	75%	89%	60%	**69%**
Use of written product spec from customers	0%	80%	65%	20%	100%	45%	33%	50%	76%	62%	53%	**58%**
Customer feedback during SPECIFICATION Development	20%	60%	80%	20%	90%	50%	40%	50%	52%	62%	60%	**58%**
Customer feedback during PRODUCT development	20%	40%	70%	20%	95%	48%	40%	57%	52%	64%	60%	**58%**
Customer feedback during PROTOTYPE / TRIAL Phase	20%	80%	60%	60%	95%	58%	67%	63%	60%	71%	67%	**66%**
Customers directly involve in NPD process	0%	60%	55%	20%	95%	38%	20%	43%	44%	47%	47%	**46%**
Customers can access to detailed design info	0%	20%	80%	0%	95%	40%	27%	23%	52%	40%	40%	**44%**
Customer have access to prototypes	0%	40%	75%	60%	95%	58%	40%	43%	72%	64%	53%	**60%**
Project teams directly contact customers	0%	80%	75%	60%	90%	63%	53%	53%	60%	67%	60%	**64%**
Simulate customers' environments in NPD	20%	60%	75%	100%	90%	73%	53%	57%	52%	78%	47%	**67%**
Involve customers in risk mitigation issue	0%	60%	55%	40%	95%	43%	20%	40%	40%	51%	60%	**48%**
Customers generally know next technology direction	0%	40%	70%	20%	75%	45%	13%	43%	68%	53%	33%	**49%**
Sell next technology or innovation direction	60%	60%	50%	60%	70%	58%	53%	67%	52%	62%	60%	**60%**
Customers have direct access to design database	0%	20%	50%	0%	95%	31%	7%	33%	24%	18%	27%	**32%**
Product specification tools: QFD	0%	100%	75%	0%	50%	25%	67%	17%	60%	67%	33%	**47%**
Six-Sigma	0%	0%	50%	0%	25%	0%	0%	33%	40%	56%	0%	**27%**
Fish-bone Diagram	0%	0%	50%	0%	25%	0%	33%	17%	0%	44%	0%	**20%**
Taguchi	0%	0%	25%	0%	25%	0%	0%	0%	0%	0%	0%	**4%**
Who determines quality standard: Industrial Standard	0%	0%	25%	0%	0%	50%	0%	17%	60%	22%	0%	**24%**
Company	100%	100%	50%	100%	0%	38%	100%	83%	60%	89%	33%	**62%**
Customer	100%	0%	100%	0%	100%	63%	33%	33%	80%	44%	67%	**60%**

- *Simulate customer environment during development phase*: Most surveyed companies claimed that they frequently studied new product performance under various conditions to ensure that the products can satisfy most customer requirements in most possible circumstances. Led by construction, defense, and medical device industries, overall application of this practice went up to 67% of all time.
- *Involve customers in risk mitigation issue*: Sometimes it is necessary to involve customers in order to reduce risks of introducing new products due to uncertainties in markets. From this survey, little less than half of surveyed companies applied this approach in NPD process. Only 48% of all time the firms engaged to this concept.
- *Customers know next innovation and technology*: This question basically asked the respondents whether their companies release information on new technology and innovations to their customers. The responses were not surprising; about half of all times firms released such information.
- *Sell next technology & innovation direction to customers*: This question succeeds the previous question to discover whether the surveyed companies offer their developed technology and innovations as a type of product to particular customers, too. The answers to this question were negatively correlated to the answers of the previous question. In other words, the less customers know the next technology, the more likely customers have to pay for the knowledge.
- *Customers have direct access to design database*: This is the least implemented practice of all approaches in the survey. On average, the surveyed firms employed this application only 32% of all time. The custom was rarely put into action especially in consumer-product industries as seen as manufacturers would not allow ordinary end-users to reach their design databases, while major industrial-product customers may have the right of entry to the databases at a certain degree.
- *Product identification-specification tools*: The survey initially suggested four famous tools: - Quality Function Deployment (QFD), Six-sigma, Fish-bone diagram, and Taguchi technique; used for capturing customer requirements and transforming into product specification. The question also allowed respondents to add tools actually used at their workplaces. The most popular tool for this purpose was QFD with 47% of respondents using it; followed by Six-sigma (27%), Fish-bone diagram (20%), and Taguchi (4%). The QFD technique ensures that customer requirements have been considered before design decisions are made. One of the most commonly reported advantages of using QFD is that it systematically reduces number of changes as a design enters production,

in some case up to 50% [Willaert et al 1998]. Additional tools responding to this question included Design for Manufacturing, Modeling & Simulation, Statistical Process Control (SPC), Multi-attribute Analysis, and some in-house systems.

- *Who determines the minimum acceptable quality level of the industry?*: This question was intended to study who: - manufacturers, customers, or third party, really have control over product quality level. The study discovered that manufacturers could define their quality standards only when they supplied commodity products such as drugs, foods, building components, medical devices, and small machineries. With more involvement in NPD process, customers had more power in made-to-order industries like ordnance, instrument, printing & publishing, and circuit board producers. Overall both sides had almost the same supremacy at about 60% while such a third-party power as industrial standard, with 24% on average, seldom dominated the quality standard. The summary of responses to this question can be illustrated as in figure 4-5.

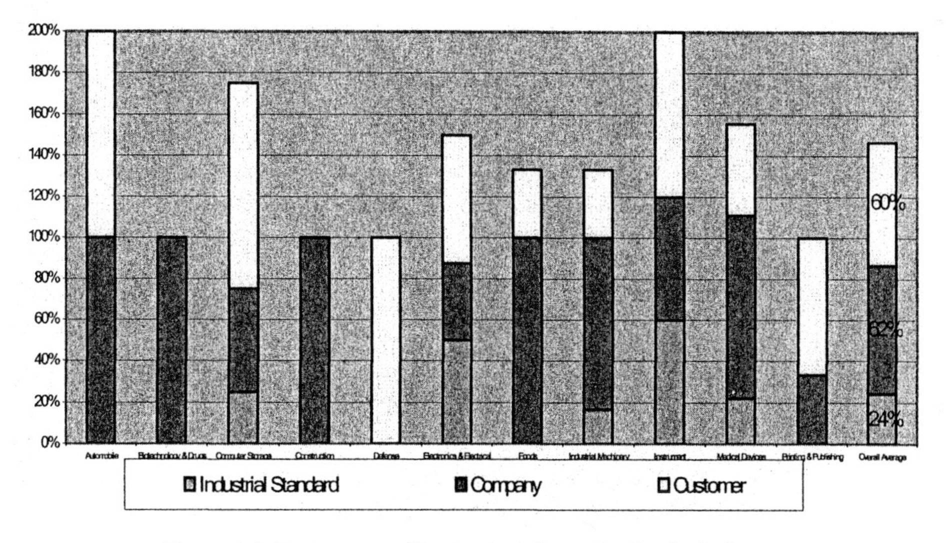

Figure 4-5. Power on quality standard determination by industry

The last question in this section asked the respondents to position their companies in a grid of market and competition environments. As seen in figure 4-6, the grid is divided to four quadrants on two axes of entry barrier and exit barrier.

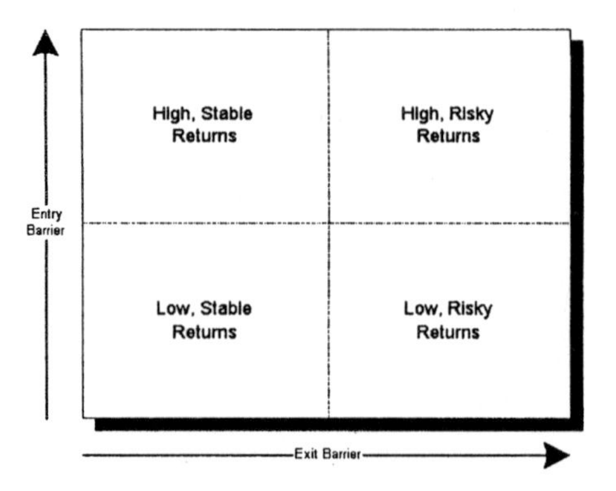

Low entry barrier, Low exit barrier => High competition, Easy to exit => Low, stable returns
High entry barrier, Low exit barrier => Low competition, Easy to exit => High, stable returns
High entry barrier, High exit barrier => Low competition, Hard to exit => High, risky returns
Low entry barrier, High exit barrier => High competition, Hard to exit => Low, risky returns

Figure 4-6. Landscape of market and competition determined by entry barrier and exit barrier

Entry barrier in business perspective means obstacles that prevent a firm to enter into a market or competition, such as high levels of investment and technology or lack of expertise available in the industry, and so forth. On contrary, exit barrier represents various reasons that foil a company in a particular competing market to quit the competition. The exit barrier may be defined as high investment that the firm has plunged into existing products or a situation that there is no better opportunity for the company to pursue a new product so that it has to continue competing in the unsatisfactory market.

For overall result, most respondents (34%) placed their companies in *low-stable returns* domain of which companies were mainly industrial-product and OEM manufacturers. There were a few food producers falling in this category, too. Table 4-6 summarizes distribution of surveyed companies on the subject of this typology..

Table 4-6. Results in counting numbers of company allocation into market-competition landscape

Industry	Low, risky returns	High, risky returns	Low, stable returns	High, stable returns
Automobile			1	
Biotechnology & Drugs	1			
Computer Storage		1	2	1
Construction			1	
Defense		1	1	2
Electronics & Electrical	2	2	3	1
Foods		1	2	
Industrial Machinery	1	1	3	1
Instrument		1	1	3
Medical Devices		5	1	2
Printing & Publishing	1	1		1
Total	5	13	15	11

High-risky returns were ranked second with 30% of respondent population. Companies within this group were largely hi-tech manufacturers in medical device, electrical, and computer industries. The third ranked class was *high-stable returns* of which members were specialists in entrenched industries like instrument, ordnance, and medical device. The high degree of technology and innovativeness of this group, besides its well-established business standard, ensured its members to enjoy their wealth for a long time. Last of all, *low-risky returns* companies wrapped up with 11% share.

4.1 New Product Development Timeliness and Schedules

Questions in this section were set to study practices and consequences of project schedule monitoring and control. NPD projects are same as normal projects that require practitioners to keep an eye on and take actions in order to keep the progress on schedule. The results are listed as follows with summary given in table 4-7:

- *Use formal schedule and/or standardized baseline schedules for NPD*: This practice was ranked among the most applicable tactics in this section with 76% of all time. When the respondents were asked in the questionnaire regarding how often their companies had provided training of the scheduling tools to NPD project teams, the responses were proportionally reduced to 64% of all time.

- *Apply project time compression technique*: Project time compression can be achieved through overlapping concurrent works. Modular approach is a popular technique that allows several modularized activities to work out simultaneously. The responses to this question were at a moderate level of 56% of all time.
- *NPD schedules are broken down into details*: This practice is expected to help NPD team leaders to precisely keep tracking project progress. However, the survey respondents found out that only 56% of all time their projects used such detailed schedules.
- *Recognize that there is a linkage between product development and an effective supply chain*: A product development process does not engage only employees in R&D, design, engineering, sales, and other functions within an organization; but rather also include outsiders like suppliers, logistics, distributors, and customers. In other words, a product development project must consider all possible aspects relating to the entire supply chain to be able to design products and processes properly. The overall responses ranked 57% of all time; much lower than expected.
- *Ability to complete projects on schedule*: This measure is supposed to be used as one of project success metrics in this research. The overall ranking was that only 54% of all NPD projects could complete in time. Only defense industry could keep their projects up to 80% on schedule.
- *The fuzzy front-end concept is recognized*: In NPD unacceptably high failure rates have often been related to insufficiencies during the early development phases e.g. idea screening, feasibility study, and business analysis. Nevertheless, only little effort is devoted to the early phases; only 6% of dollars and 16% of manpower used at these stages [Cooper, Kleinschmidt 1988]. In theory as well as in practice, project managers often indicate the front end as being one of the greatest weaknesses in a product development process. Fuzzy front-end typically involves ad-hoc decisions and ill-defined processes [Montoya et al 2000]. The term *fuzzy front-end* therefore takes root into the uncertainty and intangibility of project outcomes at the early stages of projects. Ironically, the respondents reported this concept had been recognized only 54% of all time among management and project teams.

Table 4-7. Summary of responses in NPD timeliness and schedules aspects by industry

Industry		Automobile	Biotech & Drugs	Computer Storage	Construct	Defense	Electronics & Electrical	Foods	Industrial Machinery	Instrument	Medical Devices	Printing & Publishing	Overall Average
Use formal and/or standardized baseline schedules		0%	80%	80%	100%	100%	65%	60%	73%	88%	78%	73%	**76%**
Offer training for scheduling tools		0%	80%	60%	80%	90%	55%	60%	57%	60%	75%	60%	**64%**
Apply project time compression technique		0%	60%	50%	80%	50%	53%	53%	63%	50%	71%	40%	**56%**
NPD schedules are broken down into details		0%	60%	45%	80%	90%	55%	47%	50%	52%	64%	40%	**56%**
Recognize that there is a link between PD and supply chain		40%	40%	45%	40%	55%	55%	80%	47%	56%	71%	53%	**57%**
Ability to complete projects on schedule		40%	40%	65%	20%	75%	50%	53%	60%	52%	42%	67%	**54%**
Fuzzy front-end concept is recognized		80%	40%	50%	20%	25%	65%	60%	30%	56%	73%	47%	**54%**
Modular technique is considered at the early stages of PD process		0%	40%	50%	40%	80%	60%	33%	37%	55%	47%	40%	**50%**
Customers are given tradeoffs for shorter schedules		0%	40%	45%	20%	75%	30%	33%	50%	56%	38%	47%	**43%**
Project schedule is used as a realistic tool for project tracking		0%	60%	60%	40%	95%	58%	53%	67%	68%	60%	67%	**63%**
You are often asked to commit to BEST case schedule or even tighter		60%	60%	80%	80%	95%	68%	73%	80%	84%	84%	87%	**80%**
Team or personal performance relates to schedule performance		60%	60%	60%	80%	90%	60%	53%	60%	80%	69%	67%	**67%**
Have an initiative to monitor and improve project schedules		0%	40%	60%	60%	75%	45%	53%	57%	40%	60%	47%	**53%**
Utilize collaborative engineering tools in PD efforts		0%	20%	70%	60%	80%	35%	73%	50%	40%	44%	47%	**49%**
Use baseline schedules for different types of projects		20%	40%	55%	20%	90%	45%	60%	47%	56%	44%	40%	**52%**
Scheduling tool: Microsoft Project		0%	100%	100%	100%	100%	75%	100%	83%	100%	100%	67%	**89%**
Microsoft Excel		100%	100%	75%	100%	50%	38%	33%	50%	20%	44%	67%	**49%**
Stage-Gate		0%	0%	25%	100%	50%	0%	0%	0%	20%	22%	33%	**18%**
Squeaky Wheel		0%	0%	0%	0%	25%	38%	0%	17%	40%	22%	33%	**22%**
Schedule proposed: Best case		0%	0%	25%	0%	50%	25%	0%	33%	60%	56%	33%	**36%**
Worst case		0%	0%	0%	0%	25%	13%	0%	0%	0%	0%	0%	**4%**
Realistic case		100%	100%	75%	100%	50%	50%	100%	50%	60%	33%	67%	**58%**
Schedule performance starts at:	Conceptualization	0	0	0	0	3	0	0	0	0	0	1	**9%**
Research		0	1	0	0	0	0	1	1	0	0	0	**7%**
Feasibility study		0	0	1	0	1	2	1	1	2	3	1	**27%**
Design		0	0	2	1	0	5	0	2	2	6	0	**41%**
Prototyping		1	0	1	0	0	1	1	1	1	0	1	**16%**

- *Modular techniques are considered in the early stages of product development process:* Modularity refers to the scheme by which interfaces shared among components in a given product architecture are standardized and specified to allow for greater reusability and commonality sharing of components among product families (Mikkola 2001). In other words, modular approach for NPD designs a product in such a way that its components are developed concurrently and combined with predefined interfaces that would benefit in term of shorter project schedule as mentioned in the project time compression topic. This technique requires a lot of works to do during the design phase. However, it seemed not to be popular much in the surveyed companies. Only 50% of all time were responded.
- *Customers are given tradeoff choices for shorter schedules*: In a certain NPD project, the customer may desire to get the product into market earlier than the initial plan, but such a change demands additional resources i.e. manpower, overtime charges, sub-contraction, etc. to compensate the shrunk project time. Producers may provide tradeoff choices for customers, for example, additional cost for a unit of time shortened. The responses of this question ranked the lowest score in the section with only 43% of all time application.
- *Project schedule is used as a realistic tool for project tracking*: One might find scheduling tools are merely used for showing rough schedule while most project controllers really utilize them as an effective device for project monitoring and control. This study discovered that most NPD projects really made use of schedule efficiently with 63% of all time application.
- *Project teams are often asked to commit to the best-case schedule or even tighter*: Notably, the responses to this question accounted to 80% of all time, the highest in this section. This can be implied that most project managers or above keep pushing the project teams to complete their projects within the best-case schedule, the shortest one.
- *Team or personal performance is measured as a function of schedule performance*: Generally, appraisal of project or project team members highly depends on project schedule performance. Nevertheless, this relation might have some exceptions, for instance, schedule has not been well planned or the schedule has been shrunk unrealistically. The survey results gave out the score of 67% that personal or team performance related to ability to control project schedule.
- *Have initiatives to actively monitor and decrease product development cycle time*: This question was anticipated to find out whether the

surveyed companies had any other means than schedule to keep tracking and improve product development cycle time. A little over half of all projects had employed such tools.

- *Utilize collaborative (Internet base) engineering tools*: The collaborative engineering is a systematic approach to control life cycle cost, product quality, and time to market during product development, by concurrently developing products and their related processes through information technology infrastructure (Willaert et al 1998). Especially when design teams spread across the globe, the Internet scheme is the most appropriate infrastructure for communication within limited cost and time. Regardless of highly potential benefits of collaborative engineering, the survey respondents utilized the tools only 49% of all time.
- *Use baseline schedule for different types of projects*: Baseline schedule is a sort of generic schedule that is commonly used in an organization for several projects because characteristics of products and their processes in NPD projects are similar to precedent projects. Using the baseline schedule consequently save resources and time in reinventing a new schedule for each new project. However, the survey participating companies exploited this technique only 52% of all time.
- *Scheduling tools*: There were four popular scheduling programs nominated in the survey material. However, the questionnaire also allowed respondents to add their tools into the list. Microsoft Project was the most popular tool for this purpose with 89% utilization. From the same camp, Microsoft Excel followed the leader with 49% of all projects. Squeaky Wheel Optimization, an optimization program that continuously analyzes problems in a project and iteratively reprioritizes tasks within the project with so-called greedy algorithm (Joslin, Clements 1999), was used in 22% of all surveyed firms. The least utilized tool in the proposed group was Stage-Gate, a specific program commonly used for managing new product development and product portfolio. This tool was only 18% utilized among the surveyed companies.
- *Best-case or Worst-case schedule*: Experienced projects prepare not only best-case schedules (shortest project time) but also consider relevant risks and develop the worst-case scenario schedule (with longest cycle time). Then, the projects would decide to offer the best case and/or the worst one to customers. No fixed rule best applies to the decision: - depending on project risk and competition levels. When the respondents were asked to select what type of NPD project schedules their companies

usually propose to their customers, the realistic schedule was mostly responded (58% of all firms). The timeframe of realistic schedule is somewhere in between the best-case and the worst-case scenarios. The best-case schedules had been offered for 36% while the worst-case schedules were put forward only 4% of all projects.

- *At what point in NPD cycle that schedule performance starts to be measured*: Under a normal NPD cycle, the project manager can start measuring schedule performance at conceptualization, research, feasibility study, design, or prototyping phase. Most respondents (41% of all surveyed firms) started measuring at the design phase followed by feasibility study (27%), prototyping (16%), conceptualization (9%), and research (7%).New Product Development Process

There is a common belief that a successful product development project must be achieved through a well-established process accompanied by prominent project management tools, such as, modular approach, failure mode and effect analysis (FMEA), disciplined design flow, and so forth. This section studies how manufacturers employ standard procedures and project management practices to develop their new products. The summary of the study can be viewed in Table 4-8 and results outlined below:

- *Use formal procedures for NPD*: The practitioners responded to this question that, on average, they used formal procedures for NPD at 66% of all time. Referring to learning process of the procedures, surveyed companies offered training for project members at 57% of all time.
- *Involve major suppliers and customers in NPD from the start of projects*: Most people may have realized that involving customers at the early stages of NPD projects will help reduce marketing failure rates at the later stages, which usually cost much more than rejecting unsuccessful projects at the earlier phases. However, by far fewer people recognize that incorporating suppliers to NPD projects also immensely benefits manufacturers. For example, with a combination of complementary expertise, it reduces lead times and risks. The exercise enhances flexibility, product quality, and market adaptability while reduces development costs (Chung, Kim 2003). This practice has been introduced to predominant industries in 1980s; such as, automobile, construction, electronics, and aerospace, and today it can still be seen implemented largely in these industries. The survey results positively reflected this belief with overall application of 60% of all time.
- *Use a modular approach for product design*: The modular approach in product design enables concurrent development of modularized components, which results in faster time to market. Based on a case

study of the European Automotive Manufacturer project, the distributed engineering teams, in conjunction with information technology and telecommunication systems, could save time during several stages by 10-50%. In terms of cost saving, a potential overall saving of 20% in developing time could increase sales volume by about $1.6 billion, and cut costs by about $144 million (May, Carter 2000). The overall score of this practice in this research study was 58% of all projects.

- *Use FMEA*: Failure Mode and Effects and Analysis (FMEA) is a systematic approach that identifies potential failure modes in a system, product, or manufacturing process caused by either design or manufacturing process deficiencies. It also identifies critical or significant design or process characteristics that require special controls to prevent or detect failure modes. In brief, FMEA is a tool used to prevent problems from occurring. The survey results found that FMEA had been largely used up to 62% of all projects.
- *Use risk mitigation technique*: All projects involve risk. Every time an enterprise embarks on a major project, companies need to be aware of the potential risks associated with the project. Ernst & Young, a famous global consulting company, identifies NPD risks into six different categories; business environment, information, financial, operation, transaction, and governance (Dowsett, Strydom 2002). The surveyed companies applied the technique for 57% of all time.
- *Use stage-gate technique*: Invented by Robert G Cooper, the stage-gate process (figure 4-7) is a systematic approach for product development process, divided into five phases from the preliminary assessment of an idea to its commercialization. After every stage there is a gate deciding on continuing or terminating the project. The stage-gate-model integrates the market and technological perspective. Activities are performed in parallel and decisions at the gates are made within cross-functional teams. This technique was highly applied to NPD projects among the surveyed companies with 71% for all possible projects.
- *Design processes are formally documented*: This practice was vastly implemented in the surveyed companies with 73% of all time applied.
- *Continuous learning is incorporated into project teams and company*: This question relates to organization learning concept that the knowledge and experience obtained from former projects serve as a good model for next projects. Therefore, there are two main challenging issues relating to the learning capability of projects teams: - [1] How to acquire and store knowledge effectively and [2] How to adjust behavior to reflect the

knowledge (Garvin 1997). The surveyed responses disclosed they realized this concept for 58% of all time.

- *Use a Product Data Management (PDM) system*: PDM is an information system that helps collect not only product-related data such as specifications, drawings, or bills of material, but also process-related data, for instance; design records, customer requirement reviews, and so forth. The benefits of using this system include reducing product development time due to reusability of existing designs, improving product quality and process efficiency, and ultimately enhancing NPD product and project successes. The research study revealed that 60% of all projects utilize this system.

Table 4-8. Summary of responses in NPD process issues by industry

Industry	Automobile	Biotech & Drugs	Computer Storage	Construct	Defense	Electronics & Electrical	Foods	Industrial Machinery	Instrument	Medical Devices	Printing & Publishing	Overall Average
Use well documented procedure in NPD	0%	60%	50%	20%	80%	70%	73%	67%	64%	76%	67%	**66%**
Provide training for the procedure	0%	40%	40%	0%	65%	68%	67%	50%	44%	76%	53%	**57%**
Major suppliers & customers involve in NPD from the start of projects	80%	40%	85%	80%	85%	60%	53%	53%	60%	49%	47%	**60%**
Use collaborative engineering tools in NPD	0%	40%	65%	40%	75%	38%	40%	37%	32%	31%	47%	**41%**
Use modular approach to product design	40%	60%	60%	80%	85%	50%	47%	47%	56%	58%	53%	**58%**
Use FMEA	60%	80%	75%	100%	85%	55%	33%	47%	52%	87%	20%	**62%**
Use formal risk mitigation planning technique	0%	80%	65%	80%	95%	48%	53%	37%	72%	62%	33%	**57%**
Use formal disciplined design flow technique	60%	80%	45%	80%	70%	45%	47%	33%	40%	60%	40%	**51%**
Use stage-gate technique	60%	100%	75%	100%	100%	60%	40%	50%	64%	93%	60%	**71%**
Design process are formally documented	0%	80%	65%	60%	95%	78%	47%	67%	72%	93%	47%	**73%**
Have a formal system to diagnose and correct design problem	40%	80%	45%	60%	85%	63%	47%	47%	44%	82%	40%	**61%**
Continuous learning is incorporated into project teams	100%	60%	65%	40%	80%	48%	53%	53%	52%	64%	40%	**58%**
Use PDM system	0%	40%	75%	80%	95%	50%	53%	57%	40%	76%	27%	**60%**
Consider capacity planning and resource allocation in NPD process	40%	40%	55%	80%	80%	55%	73%	50%	56%	62%	60%	**60%**
Task force team: Steering Committee	100%	100%	50%	100%	50%	50%	0%	67%	60%	67%	33%	**40%**
Peer Review	0%	0%	0%	0%	50%	0%	67%	0%	60%	22%	0%	**47%**
Phase Review	0%	100%	100%	100%	25%	13%	0%	17%	40%	56%	0%	**42%**
Expert Review	100%	100%	0%	100%	50%	25%	67%	33%	60%	56%	100%	**24%**
Accountability measures: Periodic Review	100%	100%	50%	100%	50%	50%	0%	67%	60%	67%	33%	**56%**
CPM / PERT	0%	0%	0%	0%	50%	0%	67%	0%	60%	22%	0%	**20%**
Cpk	0%	100%	100%	100%	25%	13%	0%	17%	40%	56%	0%	**36%**
Financial Performance	100%	100%	0%	100%	50%	25%	67%	33%	60%	56%	100%	**49%**
Performance measures: Time	0%	100%	50%	100%	100%	88%	0%	50%	60%	78%	67%	**67%**
Quality	100%	100%	75%	100%	25%	38%	67%	17%	60%	78%	33%	**53%**
Cost	0%	100%	50%	100%	100%	75%	100%	33%	80%	78%	67%	**71%**
Efficiency	0%	100%	25%	0%	25%	25%	33%	50%	0%	22%	67%	**29%**
Design team location: Co-located	100%	100%	75%	100%	100%	100%	100%	100%	80%	89%	67%	**91%**
Virtually-located	0%	0%	50%	0%	100%	13%	0%	17%	20%	11%	67%	**27%**

Table 4-9. Summary of responses in NPD process issues by industry (Continue)

Industry	Automobile	Biotech & Drugs	Computer Storage	Construct	Defense	Electronics & Electrical	Foods	Industrial Machinery	Instrument	Medical Devices	Printing & Publishing	Overall Average
If virtually-located: Use common design database	0	0	1	0	4	1	0	1	1	0	1	**9**
Not use common design database	0	0	1	0	0	0	0	0	0	1	1	**3**
Training sources: Internal	100%	100%	100%	100%	100%	50%	67%	83%	100%	78%	100%	**82%**
External	0%	100%	75%	100%	100%	75%	67%	83%	80%	78%	67%	**78%**
Supplier	100%	100%	0%	0%	50%	13%	67%	33%	20%	11%	67%	**29%**
Industrial Conferences	0%	100%	50%	0%	100%	63%	100%	33%	60%	33%	67%	**56%**
Average NPD project team size	2	8	24.25	30	119.50	6.75	25.33	5.20	12.80	17.67	7.67	**23.65**
Formal process imple time: < 1 year	0	0	1	0	0	0	0	0	1	0	1	**3**
1-5 years	0	1	2	1	1	4	1	3	2	3	2	**20**
5-10 years	0	0	1	0	2	2	0	2	2	5	0	**14**
> 10 years	0	0	0	0	1	1	1	0	0	1	0	**4**
Communication tools: Internet	100%	100%	100%	100%	75%	63%	100%	50%	80%	67%	67%	**73%**
Intranet	0%	0%	25%	0%	100%	75%	100%	17%	40%	89%	100%	**62%**
Satellite	0%	0%	0%	0%	25%	0%	0%	0%	0%	11%	0%	**4%**
Teleconference	0%	100%	75%	100%	100%	100%	67%	33%	100%	78%	67%	**78%**
VPN	0%	0%	0%	0%	100%	0%	0%	17%	0%	0%	33%	**13%**
MS NetMeeting	0%	0%	50%	0%	75%	13%	67%	33%	0%	22%	33%	**29%**

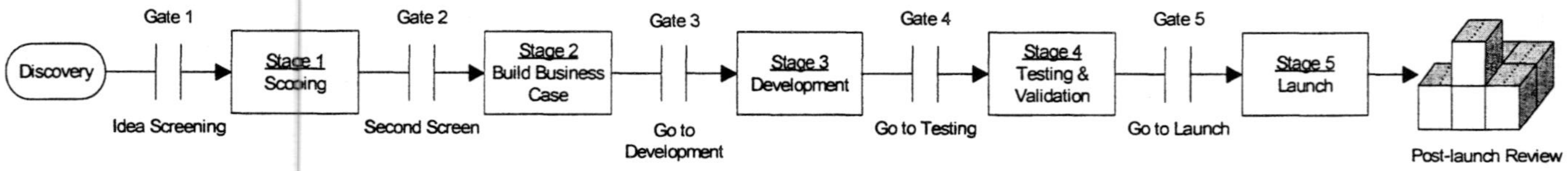

Figure 4-7. Model of stage-gate process along with discovery and post-launch review stages (Cooper et al 2002)

- Consider capacity planning and resource allocation: Both capacity planning
- and resource allocation are crucial practices in all project management especially for an organization having concurrent, on-going projects. The challenges of managing capacity and resources are how to balance fluctuating needs and prioritize concurrent demands. The survey concluded this practice at 60% utilization.
- Task force team to crosscheck and/or improve projects: More than a half of surveyed respondents did not think this group of task-force team was necessary for product or process success. Overall, the proposed teams were ranked as 40%, 47%, 42%, and 24% for steering committee, peer review, phase review, and expert review respectively. See Figure 4-8.

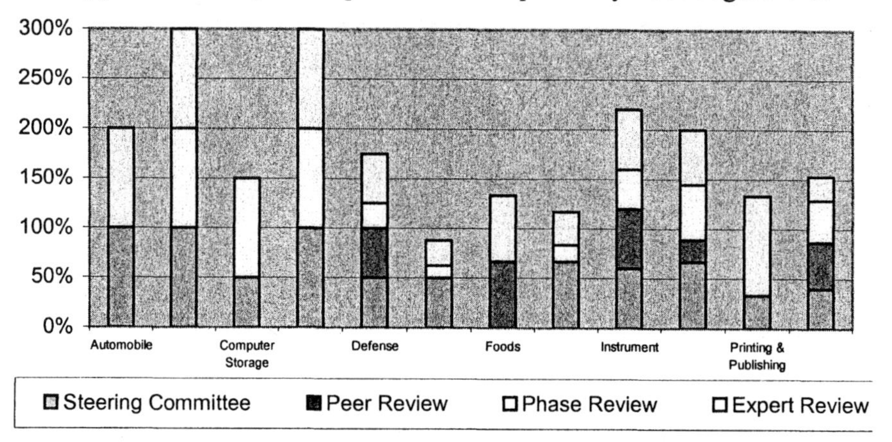

Figure 4-8. Task force teams to crosscheck and improve NPD projects

- *Accountability measures to assure product success*: Based on the nominated measures in the questionnaire, the surveyed companies claimed they used periodic review (56% of all companies), financial monitoring and control (49%), capability index, C_{pk} (36%), and critical path method (CPM) and project evaluation and review technique (20%). See Figure 4-9.

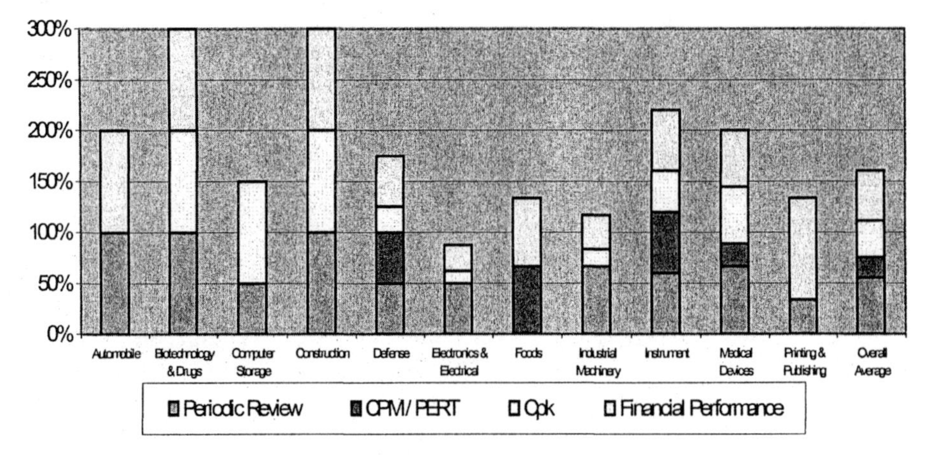

Figure 4-9. Accountability measures to assure product success

- *Metrics to measure project performance*: Cost came up to be number one on the rank with 71% of all companies followed by time (67%), quality (53%), and finally efficiency (29%). See figure 4-10.

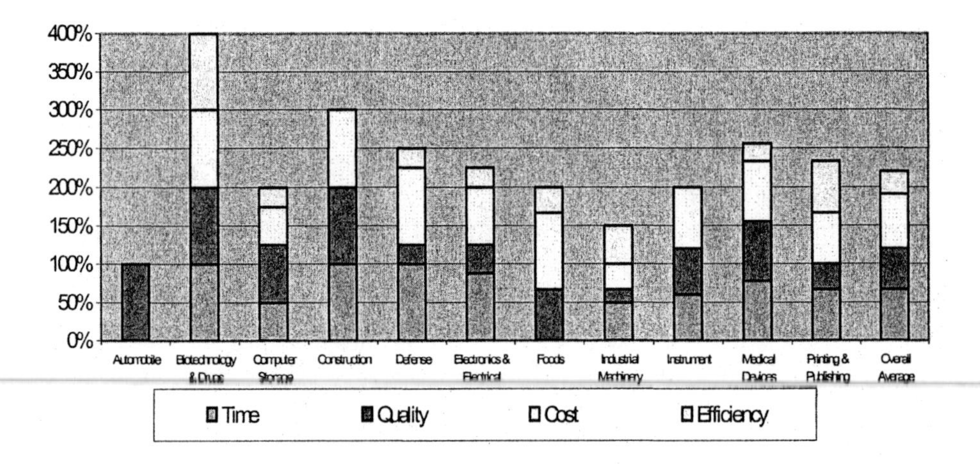

Figure 4-10. Metrics to measure NPD project performance

- *Design team location*: When the respondents were asked to define whether their design teams were co-located, virtually-located, or both; the results are shown in Figure 4-11. 73% of surveyed firms stated they were purely co-located, while 9% were solely virtually located, and 18% had both types of locations. Remarkably, the design teams in defense and printing & publishing industries had very high ratios of the virtual teams. The location of the team is an important factor to improve communication. At General Motors, design engineers and manufacturing engineers work in the same office, usually within 10 feet of each other. At AT&T, there is a 50-yards rule that means that the probability of communication among team members decreases by 80% when team members are more than 50 yards apart. To improve communication of geographically dispersed teams, they must be supported with extensive uses of electronic mail (e-mail), mobile phone, videoconference, or Internet-based groupware (Willaert et al 1998).

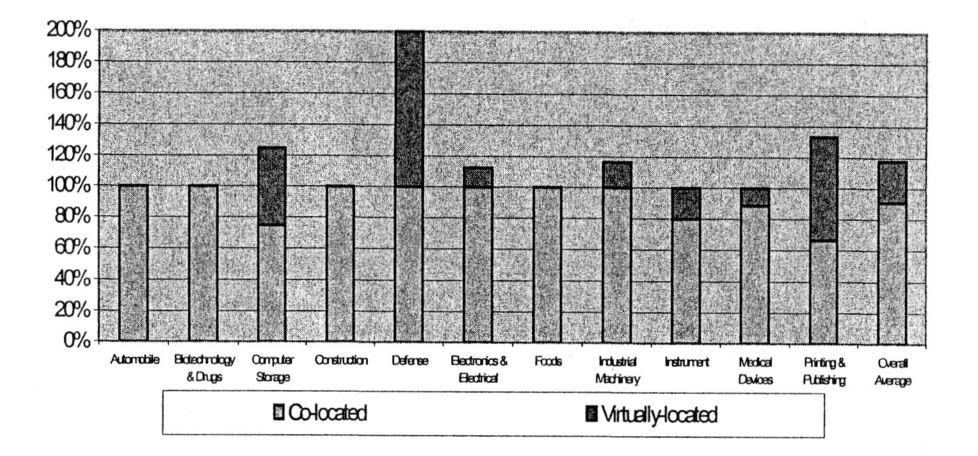

Figure 4-11. Design team locations

- *Training sources*: It is crucial to keep knowledge and skills of NPD project teams up to date with the industry. Most companies offer educational opportunities for their employees either in-house training, advanced degree support, or attending industrial conferences. Referring to Figure 4-12, 82% of surveyed companies provided internal trainings, followed by 78% on external trainings, 56% on industrial conferences, and 29% on supplier trainings.

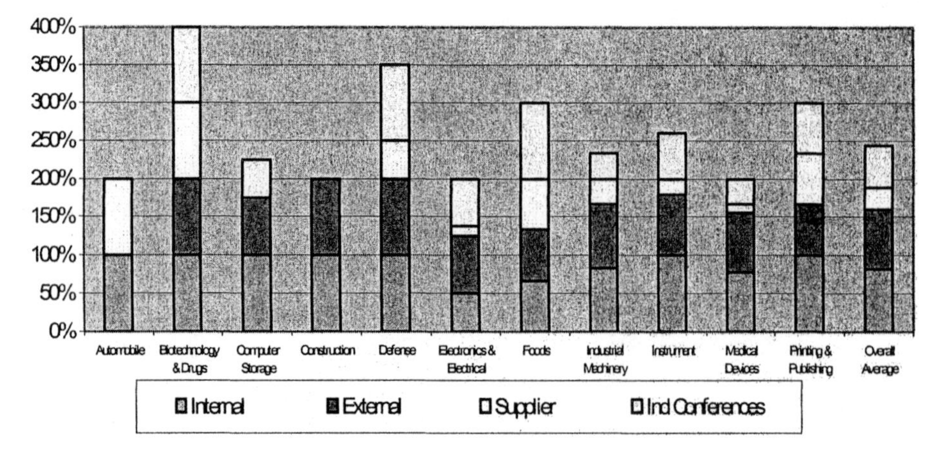

Figure 4-12. Training sources for NPD project teams

- *Period of implementing a standardized NPD procedure*: This question was meant to study for how long companies have implemented formal NPD processes. Almost half of responses have been familiar with the formal processes for 1-5 years. Another 34% of surveyed firms have utilized their processes for 5-10 years. Only 10% have used the processes for more than 10 years. For the smallest portion of 7%, the companies have used the processes for less than a year. Graphical illustration of this issue is shown in Figure 4-13.

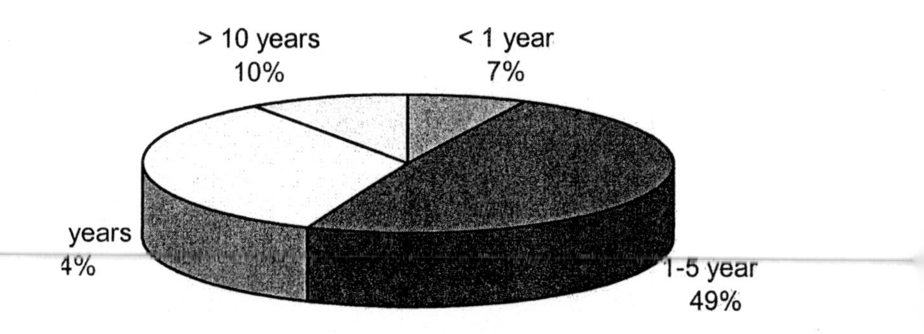

Figure 4-13. Period of implementing standardized NPD processes of surveyed companies

- Communication tools: Communication tools are a vital factor for a good coordination among remote teams as well as customers and suppliers. Today's information technology and telecommunication (IT&T); based on heterogeneous hardware, software platforms, and networks, offer the potential for more effective and efficient synchronous and asynchronous communication and therefore better collaboration between virtual team members (May, Carter 2001). 78% of the studied companies utilized videoconference for telecommunication, followed closely by the Internet-based application e.g. e-mail and web surfing for 73%. The internal networking system like Intranet was also popularly used for 62%. The summary of the tools can be seen as in Figure 4-14.

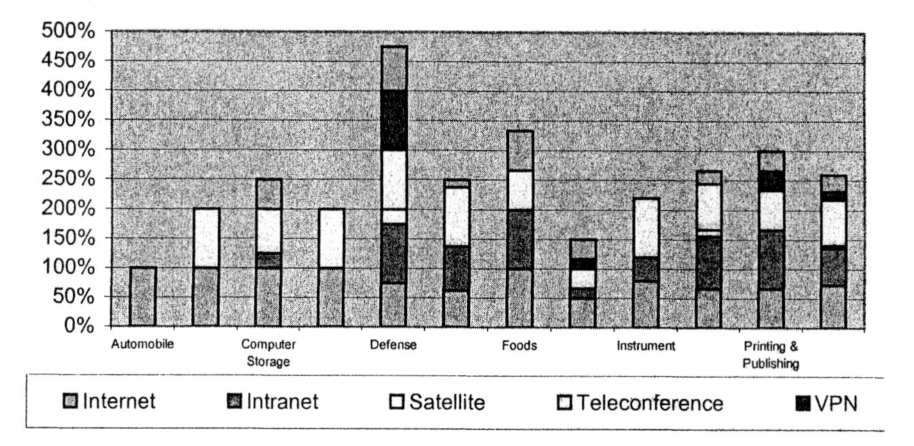

Figure 4-14. Communication tools used across the NPD project teams as well as customers and suppliers

5. NEW PRODUCT RESEARCH

Directly related to innovativeness and long-term competency, research is one of crucial activities in the NPD process. Number of organizations segregate this activity from product development and anticipate outcomes out of this function as new technologies or new platforms for further development into commercialized products in another department. Nonetheless, there are also many companies combining both functions under one roof of research and development (R&D). Many studies confirm the linkage between R&D efforts at the firm level and increased performance and competitive competency. Still, an increase by a firm of its R&D budget cannot certainly assure a proportional growth in inovative output (Conceição et al 2002).

Budgeting R&D is crucial for at least two reasons: if too much is spent, short-term financial stability is at risk, while, if the budget is too small, long-term competitiveness is threatened (Heidenberger et al 2003). Based on a research of Strategic Directions International Inc., U.S. R&D spending increased 2.4% in 2002 to $291.7 billion; equivalent to 2.79% of Gross Domestic Products (GDP). Industry accounted for 66% of R&D investments in 2002, while the U.S. government accounted for 28%. However, in real terms, industry spending decreased 1%, government spending increased 10%, and spending from other sources (including academia) rose 7% (Strategic Directions International Inc. 2003). Across the Atlantic Ocean, current EU average R&D spending topped only 1.9% of GDP. Despite economic slowdown, EU has tried to boost the budget to 3% level in 2003 (European Report 2003).

The questions in this section aim to discover the practices connecting to the research activity. The results of this section are summarized by industry in Table 4-10

- *Use feasibility studies for areas lacking of expertise*: New technology platforms or areas without expertise are critical for a company to decide whether to pursue them or not. The firm thus has to carefully study the possibility and feasibility of employing such new technologies. In overall view, the studied companies adopted this practice only 54% of all time.
- *Define a specific research phase in NPD projects*: It is not necessary to separate the research activities to a specific phase in all NPD projects; especially for small projects that may need to develop new products from incremental innovation only. On average, each surveyed companies defined a specific research phase in their NPD projects at a ratio of 59%.
- *Assign a separate group of employees for research activities*: With the same reason as mentioned at the previous question, this practice was applied at a rate of 66% of all time.
- *Do most of research activities during the NPD cycle*: This question was anticipated to look for the ratio of research activities being conducted in NPD process. Some companies may have a separate function working specifically to innovate only new technologies and product platforms but not develop into commercialized products. Such a research center can also make profit by selling technologies and associated knowledge to both internal and external customers. Research centers of Xerox, Motorola, and Intel serve as good examples for this kind of profit center.

The respondents disclosed that they used, on average, 64% of NPD cycle time on research, which similar to Willaert's study that engineers usually spend 1/3 of their time on designing but exploit the rest of time for researching to verify their design data (Willaert et al 1998).

- *Product innovations are part of the research activities*: Basically, most people believe that company's product innovations are always the result of research activities, but it is not always true. As mentioned in chapter II, the incremental and imitative innovations can be easily adopted from customers, suppliers, or even competitors. The survey results confirmed this phenomenon that only 67% of all product innovations in surveyed companies were the result of their research activities.

Table 4-10. Summary of responses on new product research by industry

Industry	Automobile	Biotech & Drugs	Computer Storage	Construct	Defense	Electronics & Electrical	Foods	Industrial Machinery	Instrument	Medical Devices	Printing & Publishing	Overall Average
Use feasibility study for areas lacking of expertise	0%	100%	40%	100%	67%	50%	80%	27%	40%	78%	33%	**54%**
Define research phase in NPD projects	0%	80%	60%	80%	85%	60%	53%	27%	44%	82%	47%	**59%**
Assign a separate group of employees for research activities	0%	80%	65%	100%	90%	48%	73%	67%	68%	78%	40%	**66%**
Do most of research activities in the NPD cycle	80%	80%	60%	20%	60%	63%	80%	67%	52%	64%	73%	**64%**
Limit the amount of risk/ unknown	80%	60%	70%	80%	80%	63%	53%	57%	52%	73%	80%	**66%**
Product innovations are part of research activities	20%	60%	75%	80%	75%	65%	80%	53%	56%	78%	60%	**67%**
Internal Expertise	0%	100%	100%	100%	100%	88%	100%	83%	100%	100%	100%	**93%**
External Expertise	100%	0%	0%	100%	100%	63%	33%	33%	80%	44%	0%	**49%**
Market risk is higher	0	1	1	0	0	5	2	1	2	2	1	**15**
Technical risk is higher	0	0	2	0	3	2	1	0	2	4	0	**14**
Both are the same level	1	0	1	1	0	1	1	4	1	3	2	**15**

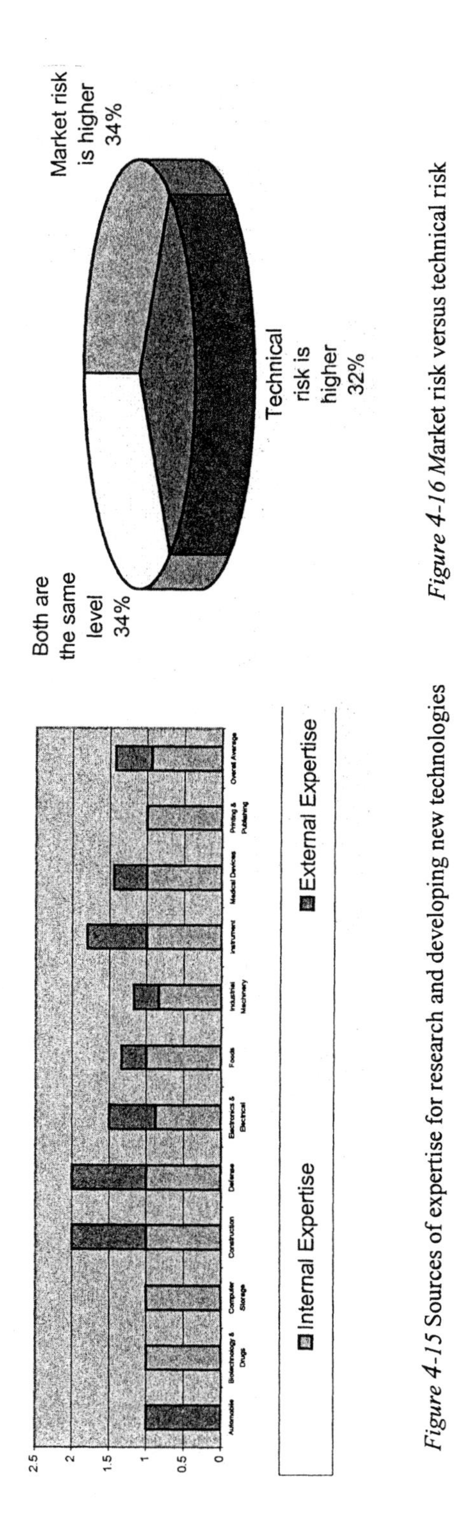

Figure 4-15 Sources of expertise for research and developing new technologies

Figure 4-16 Market risk versus technical risk

- *Sources of expertise for research and developing new technologies*: Sometimes, good ideas for product innovations come from external sources such as customers, suppliers, competitors, and consultants. This question asked respondents to identify their sources of knowledge, which can be internal and/or external expertise. As shown in figure 4-15, the overall responding companies utilized internal expertise for 93% and external expertise for 49% for all projects. While the companies from biotechnology, computer storage, and printing industries fully used only internal expertise, the company from automobile industry utilized knowledge only from external source.
- *Marketing risk versus technical risk*: This question was intended to study how important the marketing risk was recognized as opposed to technical risk in NPD projects. The results, as seen in figure 4-16, show that the studied group of NPD practitioners equally valued the significance of both marketing and technical risks.

6. TEAMWORK AND LEADERSHIP

Similar to any other types of projects, a successful NPD project needs a good cooperative teamwork and visionary leadership. The Product Development Management Association (PDMA)'s researches on NPD practice in 1991 and 1995 revealed that the more successful firms in NPD tended to use multi-disciplinary product development teams more than the less successful ones. In addition, the studies also found that NPD is melded into firm's organization through multiple structures in each firm (Griffin 1997; Page 1993; Little 1991). Another factor producing successful NPD is the need for tangible and visible top management support of NPD, especially in terms of providing funding and resources and consistent strategies. On the other hand, the top management should not so much involve in details that prevents the project teams from having flexibility, creativity freedom, and participative decisions (Bonner et al 2002). Based on the summary in table 4-11, the survey results are discussed as follows.

- *Have a formal project management role/title*: The question was aimed to study how formally companies assign project management roles to NPD project leaders. The study found the surveyed companies broadly employed this practice at a ratio of 72% of all time.
- *Use program mentoring*: Some companies provide trainings, tools, and networks to support mentoring for projects. The mentoring program assists project members who never have experience in NPD by

identifying and recruiting mentors to meet requirements of individuals. The survey result showed that this practice was not well implemented in the companies surveyed since they used this technique only 42% in aggregate. Furthermore, there was no single industry applying this practice for more than 60% of all time.

Table 4-11. Summary of responses on teamwork and leadership issues by industry

Industry	Automobile	Biotech & Drugs	Computer Storage	Construct	Defense	Electronics & Electrical	Foods	Industrial Machinery	Instrument	Medical Devices	Printing & Publishing	Overall Average
Have formal project management roles/titles	0%	100%	80%	80%	90%	73%	67%	60%	80%	67%	73%	**72%**
Use Program Mentoring	0%	40%	45%	20%	60%	30%	53%	40%	30%	51%	47%	**42%**
Use Program Sponsorship	0%	40%	45%	20%	55%	40%	53%	43%	30%	45%	60%	**43%**
Use Cross-functional Teams in Product Development	80%	80%	90%	80%	100%	73%	80%	63%	76%	89%	87%	**81%**
Executive Actively Involve in NPD	100%	60%	60%	60%	75%	53%	60%	70%	68%	78%	67%	**67%**
Engineers are Responsible/ Accountable for Project Outcomes	20%	80%	85%	80%	85%	70%	60%	80%	80%	82%	67%	**76%**
Number of Project/ Program Managers	0	5	4	2	48.75	6.71	7.33	6	13.4	19.67	12.67	**13.93**
Sources of new product ideas: Executive	100%	100%	0%	100%	25%	63%	67%	0%	60%	44%	100%	**47%**
Employee	100%	100%	100%	100%	50%	75%	67%	83%	100%	89%	100%	**84%**
Customer	100%	100%	75%	100%	100%	100%	67%	67%	80%	78%	100%	**84%**
Supplier	100%	0%	25%	0%	25%	0%	33%	0%	0%	11%	0%	**11%**
Member-leader selection criteria: Skill	0%	100%	75%	0%	100%	63%	67%	50%	40%	78%	100%	**67%**
Seniority	0%	0%	25%	0%	0%	13%	0%	0%	40%	33%	67%	**20%**
Availability	100%	100%	50%	0%	25%	63%	33%	50%	80%	56%	67%	**56%**
Rewarding for project success: Recognition	0%	100%	75%	0%	100%	63%	100%	67%	80%	56%	67%	**69%**
Financial rewards	0%	100%	25%	0%	75%	25%	33%	17%	20%	67%	0%	**36%**
Non-financial rewards	0%	100%	50%	0%	25%	0%	0%	0%	40%	33%	0%	**20%**

- *Use program sponsorship*: The program sponsorship can be viewed in two different perspectives: being a company that gives or gains technologies. The first type means the company has been sponsored with an amount of funds to innovate or create something for the sponsor. On the other hand, if the company sponsor other research centers, either private, governmental, or university facility, to study something for it, it would fall into the second type. Similar to mentoring, this approach was not well exercised among surveyed companies. Only 43% of all time was reported in the survey result.
- *Use cross-functional team in NPD process*: Similar to other studies, most companies in the study used the cross-functional teams extensively with 81% of all time.
- *Executive actively involve in NPD*: As mentioned earlier, successful NPD projects require executive to actively support in strategic level but not too much in operational level. The study showed that overall companies had management active involvement around 67% of all time (table 4-11). The study on behaviors of private and public companies in this subject is interesting as shown in table 4-12. With by far fewer employees, the private companies surveyed had lower degree of management involvement in NPD process than the public counterparts as one can see from mean and median values in the table. The level of management involvement can reflect the average product success level.

Table 4-12. Levels of executive involvement and consequences in private and public companies

Ownership	Results on Executive Involvement			Average Employee	Average Product Success
	Mean	S.D.	Median		
Private companies	57.14%	24.63%	60%	1,177	53.85%
Public companies	71.61%	22.38%	80%	20,195	62.71%
Overall	67.11%	23.80%	60%	14,278	60.09%

- *Sources of new product ideas*: Many new product ideas are not initiated only by the direct functions, but rather by all possible sources in the supply chain; such as, executive, all levels and functions of employees, customers, and suppliers. The survey discovered that the most ideas were equally generated from employees and customers at a ratio of 84% of all time. Top management had a role on this issue only 47% of all ideas. Customers were the least contributing source of the ideas in this study (11%). However, the manners of this issue varied from industry to industry, as seen in figure 4-17.

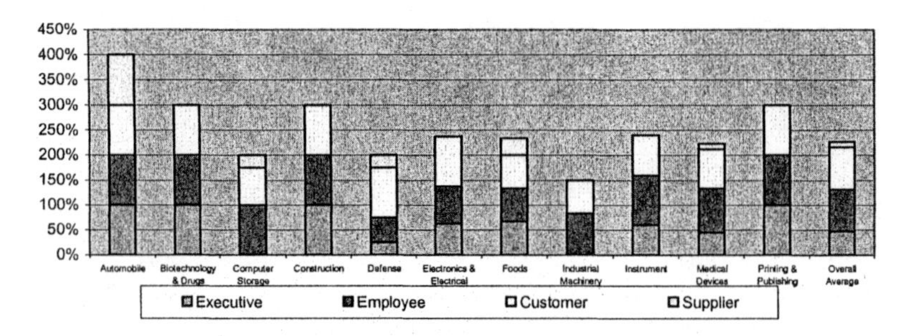

Figure 4-77. Sources of new product ideas by industry

- *Factors used to select NPD project teams and leaders*: The study results showed that the NPD teams were formed and the leaders were essentially selected from skills appropriate for individual projects (with 67% of all time). Availability was secondly ranked with 56%. The surveyed firms used seniority as the least means to select the team members or leaders with only 20%. The preferred factors vary by industry as depicted in figure 4-18.

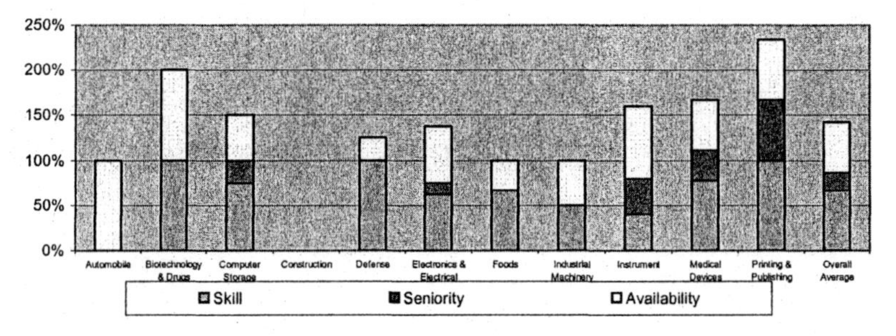

Figure 4-18. Factors used to select NPD project team members and leaders

- *Rewards for successful projects*: One of the popular practices used for recompensing an accomplishment of an NPD project team as well as encouraging other teams to complete their project successfully is rewarding system. There are numbers of ways to reward teamwork; such as recognition; financial rewards i.e. bonus, profit sharing, etc.; and non-financial rewards i.e. project dinner, extra vacation, and so forth. The study uncovered that surveyed companies had rewarded project teams mostly in the form of recognition with 67% of all time. While the financial rewards were utilized for 36%, the non-financial rewards were used only 20% of all time. The graphic of this practice is shown in figure 4-19.

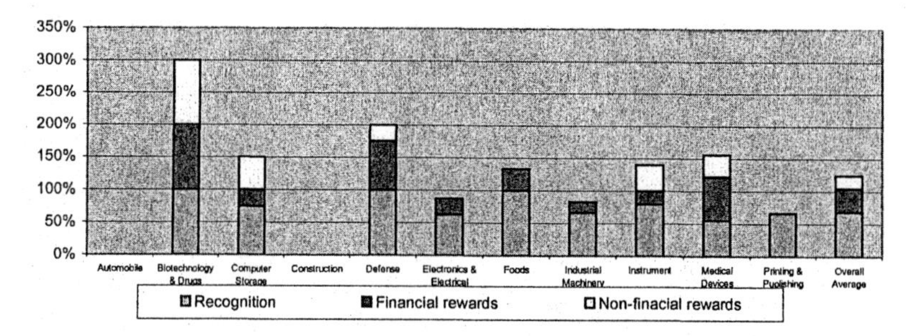

Figure 4-19. Rewards for meeting project goals

- *Number of project managers*: In a company, the number of project managers relates, but not in proportion, to number of projects at a certain time. The number also depends very much on industry type, size of the company, NPD organization structure, and rate of new products to be launched. Referring to figure 4-20, the average number of project managers in defense industry greatly outnumbered those of other industries. In average, the surveyed companies had up to 14 project managers.

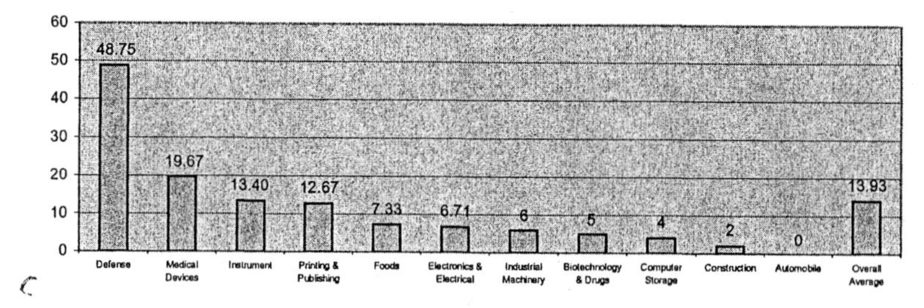

Figure 4-20. Average number of project managers in each company by industry

7. HUMAN RESOURCE DEVELOPMENT

NPD success today heavily relies on how much project members can learn and adopt good practices from their own experience and other successful stories. Therefore, a systematic training approach for NPD project team members is crucial for two reasons; first is to brush up the employees to have adequate knowledge for developing products, and second one is to retain those employees within the organization. In order to systematically design an effective training program for an employee, it is important to include following steps: - assessing training needs, training gap analysis, training conduction, monitoring, and evaluation (Goldstein, Ford 2002).

This survey section studies how companies handle the training processes in order to develop their human resources to be ready for NPD works. Ironically, the companies in survey did not pay much attention to the front end of training management. However, they highly supported learning process of employees no matter whether they worked for NPD projects. The summary of results is shown in table 4-13.

- *Formally assess internal expertise and training needs for employees*: The first step of training management is the assessment of internal expertise and training needs for employees. This can be done through surveys asking both project managers and team members for the types and levels of knowledge needed to complete a particular project. The outcomes of the study are the training needs. In reality, the survey results suggested that the surveyed companies assess the needs only 42% of all time.
- *Have formal training gap analyses*: When the needs are enlisted, the next step is to compare existing qualifications of team members to the required levels of expertise. The discrepancies will be the training gaps. From this stage, a training department or human resource development (HRD) department can take the training gaps to design a proper training program for each individual. This practice again was not well applied much among the surveyed companies. They devoted only 36% of all training opportunities to this approach.
- *Actively monitor training needs and status*: Typically, after the training programs have been carried out, it is co-responsibility of the project managers and HRD to continuously monitor the training needs whether they are fully fulfilled or need adjustment. The surveyed respondents applied this practice only 47% of all time.

- *Have a minimum set of training for employees before working with NPD*: This could be a good prerequisite standard for any employee who is going to join an NPD project to achieve before being admitted to work. However, one might argue that, in some cases, there is not much time to fully prepare staffs, who are urgently drawn from multi-functions, before projects start. One must realize that the project performance might be jeopardized if its members do not understand its process or product at all. Therefore, there should be at least a minimum set of necessary knowledge, which is common for any new product projects in an organization, available for a new comer to a project. The surveyed companies responded to this approach at only 41% of all time.
- *Company pays for seminars, trainings, tuition for qualified employees*: This is the only practice that companies in study commonly exercised extensively. Most companies were generous to support learning needs of employees, accounted to 84% of all time. The support could be ranging from funding for in-house and external training programs to give-away scholarships for advanced degrees.

8. TECHNOLOGY DEPLOYMENT

In a new product development process, selecting an appropriate technology can determine efficiency and effectiveness of the process. Technology deployment is a process to bring technology from development to operation. It starts with planning a product roadmap, a series of products underway to develop and launch to market within next 5-10 years, for example. The next step is to design another roadmap for technologies supporting products, process, and other systems. The technology deployment can be managed through these four steps: - Technology forecasting, selection, transferring, and termination. Summary of survey results by industry are presented in table 4-14.

- *Consider technology deployment a key factor for NPD success*: Most survey respondents believed that deploying technology properly plays a great part to successful NPD. In aggregate, the response rate was up to 70% of all time.
- *Develop a product cycle plan or a product roadmap*: Since the technologies to be deployed must conform to products to be developed, first a company must establish a product cycle plan, also known as a product roadmap. Not only be useful for technology planning, the product cycle plan can also be used for planning resources and funding at

the same time. On average, the surveyed firms developed their product roadmaps for 58% of all time.

- *Product development cycle closely follows the product roadmap*: Once the companies have their master plans for future products, it is interesting to know how well they can follow the plans. The average of responses to this question, after omitting irrelevant answers, was about 56% of all time.
- *Develop a technology roadmap for product, process, and other systems*: The next step in planning technology deployment is to define related technologies aimed to support the products as well as their processes and systems; such as automation, information systems, and product data management systems.
- *Deploy technology management*: Responding to the question of how often the companies plan their technology deployment as a whole, the respondents disclosed that only 48% of all time they performed the practice.

Table 4-12. Summary of responses on human resource development issues by industry

Industry	Automobile	Biotech & Drugs	Computer Storage	Construct	Defense	Electronics & Electrical	Foods	Industrial Machinery	Instrument	Medical Devices	Printing & Publishing	Overall Average
Formally assess internal expertise + training needs	0%	60%	35%	20%	70%	30%	60%	37%	24%	56%	40%	**42%**
Have formal training gap analysis	0%	60%	50%	40%	70%	33%	53%	27%	28%	31%	20%	**36%**
Actively monitor training needs and status	0%	80%	45%	40%	90%	43%	40%	40%	32%	58%	27%	**47%**
Have training / job readiness standards	0%	80%	55%	20%	70%	40%	53%	50%	56%	51%	33%	**49%**
Have a minimum set of training for employees before working with NPD	0%	100%	40%	20%	80%	30%	53%	33%	24%	44%	40%	**41%**
Company pays for seminars/ trainings/ tuition for qualified employees	40%	80%	100%	100%	100%	85%	93%	77%	68%	89%	73%	**84%**

Table 4-13. Summary of responses on technology deployment issues by industry

Industry	Automobile	Biotech & Drugs	Computer Storage	Construct	Defense	Electronics & Electrical	Foods	Industrial Machinery	Instrument	Medical Devices	Printing & Publishing	Overall Average
Consider Technology Deployment a key factor for NPD success	40%	60%	75%	80%	80%	60%	60%	63%	76%	82%	60%	**70%**
Develop a Product Cycle Plan or a Product Roadmap	0%	40%	70%	100%	65%	48%	60%	43%	52%	76%	53%	**58%**
Your product development closely follows the Product Roadmap	-	40%	55%	60%	40%	48%	47%	40%	52%	73%	53%	**56%**
Develop a Technology Roadmap for product, process, and other systems	0%	60%	70%	80%	75%	40%	73%	40%	60%	64%	60%	**57%**
Deploy Technology Management	0%	60%	60%	60%	70%	30%	80%	30%	40%	58%	53%	**48%**

Table 4-14. Summary of responses on cost / profit margin / return issues by industry

Industry	Automobile	Biotech & Drugs	Computer Storage	Construct	Defense	Electronics & Electrical	Foods	Industrial Machinery	Instrument	Medical Devices	Printing & Publishing	Overall Average
Recalculate ROI+IRR when program schedule slips	0%	60%	35%	80%	40%	30%	47%	50%	40%	58%	47%	**44%**
Projects have ever been rejected if the financial return is not adequate	100%	40%	45%	60%	40%	50%	73%	60%	40%	69%	80%	**58%**
Use hurdle rate, min ROI, min IRR to determine whether to pursue an NPD	0%	40%	60%	100%	40%	55%	73%	33%	56%	76%	60%	**57%**
Actual ROI+IRR are measured after the product is put into production	0%	40%	45%	40%	40%	48%	67%	27%	32%	53%	60%	**44%**
Your products are labor intensive	40%	60%	60%	80%	80%	53%	53%	50%	68%	69%	80%	**63%**
Have a minimum acceptable profit margin on new products	0%	80%	50%	80%	47%	80%	47%	43%	56%	71%	80%	**61%**
PRODUCT costs are monitored during the development process	60%	40%	55%	40%	75%	68%	60%	67%	64%	78%	67%	**67%**
PROJECT costs are monitored during the development process	60%	80%	55%	80%	95%	60%	53%	50%	60%	71%	60%	**64%**
Project capital costs are scheduled prior to NPD launches	100%	80%	65%	80%	75%	70%	67%	80%	52%	84%	73%	**73%**
Monitor capital costs, capital at risk, impaired capital, and capital usage	0%	60%	45%	80%	75%	45%	67%	57%	32%	71%	53%	**55%**
Have ever delayed PD projects due to inadequate cash flow for capital items	100%	60%	25%	20%	33%	18%	67%	47%	36%	33%	47%	**37%**
Monitor capital outlay to cash flow requirements	0%	60%	40%	60%	70%	28%	67%	47%	55%	56%	53%	**49%**
Use Activity-based Costing and Analysis	0%	60%	20%	0%	45%	25%	93%	32%	16%	49%	53%	**37%**
Compute and track product life cycle cost based on reliability data	0%	80%	35%	20%	80%	23%	60%	32%	32%	49%	47%	**42%**
Cost is the issue	1	0	0	1	0	4	2	4	0	3	1	**16**
Products are one-of-a-kind	0	0	0	0	1	0	1	0	1	0	0	**3**
Cost issue varies by products	0	1	4	0	3	4	0	2	4	6	2	**26**

9. COST / PROFIT MARGIN / RETURN

The questions in this section were intended to explore financial evaluation techniques generally used for monitoring and controlling project costs and return opportunities in ordinary projects. In aggregate, the companies usually used the financial indicators for forecasting the possible outcomes of projects and monitoring costs during the project period, but rarely used to recheck the actual outcomes at postmortem. The summary of responses by industry is illustrated in table 4-15.

- *Recalculate ROI[3] and IRR[4] when program schedule slips*: Once a project schedule slips, more expenses will incur especially from interests that grow over time. The pre-calculated return on investment, in terms of ROI and IRR, should be recalculated to reflect the changes on the investment side. The respondents revealed that they, on average, recalculated the economic indices only 44% of all time.
- *Projects have ever been rejected if the financial return is not adequate*: The NPD project practitioners followed this tactic 58% of all time.
- *Use hurdle rate[5], min ROI, min IRR for determining whether to pursue an NPD*: The hurdle rate is a minimum rate of return that a project requires to be considered viable for investment. This question was aimed to find out if the project teams had considered the financial opportunities of their projects, in terms of ROI and IRR in oppose to hurdle rate, prior to the project launches. The results were 57% in average.
- *Actual ROI and IRR are measured after a product is put into production to determine its success*: In most cases, the ROI and IRR are obtained by using estimated incomes, expenses, and interest rates to calculate in formulas. To determine project and product success, the project manager

[3] Return on investment (ROI) measures the financial benefits of an activity against its associated costs. Depending on the application of the ROI equation, the benefits may be referred to as future income, earnings, or profits and the associated costs or invested capital may be described as the total project costs. Equation is: ROI = (Profits / Invested capital) * 100%

[4] Internal Rate of Return (IRR) is the discount rate which makes the net present value of a project equal to zero. It is usually used to determine if a project should be accepted or not. An IRR rate higher than the cost of capital implies that the project should be approved since the company yields economic gains in excess of its cost of capital. The IRR method is popular in some companies as an alternative to ROI or payback period calculations.

[5] In budgeting for capital expenditures, hurdle rate is the rate of return that must be exceeded for the investment to be viable. Above the hurdle rate, the investment should be made; but below the hurdle rate, the investment should not be made.

should perform a postmortem analysis by taking the real financial figures to compute actual ROI and IRR. The surveyed companies implemented this approach only 44% of all time.

- *Have a minimum acceptable profit margin on new products*: During the feasibility study stage in NPD cycle, some companies might estimate a profit margin for a particular new product. The number is used for determining whether to pursue the project by comparing to a minimum acceptable profit margin. The companies on average performed like this for 61% of all time.
- *Product costs are monitored during the development process*: The responding companies monitored product costs for 67% of all time.
- *Project costs are monitored during the development process*: The project costs were also monitored during the NPD process for 64% of all time.
- *Project capital costs are scheduled prior to NPD launches*: The project capital costs were on average 73% estimated prior to project launches, especially during the business analysis period
- *Continuously monitor capital costs, capital at risk, impaired capital[6], and capital usage*: The responding companies applied this practice on average 55% of all time.
- Have ever delayed a product development project due to the inadequate cash flow: The respondents experienced such an event at a low rate of 37%.
- *Monitor capital outlay to cash flow requirements*: In a project, outlaying money for capital items must be carried out in regard to cash flow capability. The surveyed companied exercised this manner 49% of all time.
- *Use Activity Based Costing and Analysis*: Activity based costing[7] is an accounting methodology that assigns costs to activities rather than products or services. This enables resource and overhead costs to be more accurately assigned to the products and the services that consume them. The surveyed companied used this tools at a very low level of 37% on average.

[6] The capital of a company is said to be impaired if its liabilities, subtracted from its assets, leave less than the stated amount of capital.

[7] Activity based costing is an accounting methodology that assigns costs to activities rather than products or services. This enables resource and overhead costs to be more accurately assigned to the products and the services that consume them.

- *Compute and track product life cycle costs based on reliability data*: In the context of NPD, product life cycle costs include not only expenses incurred during the NPD process, but also other expenses occurring after the product has been put into production and market. For example, there might be some modification of the design of the launched product that requires a project engineer to continue working with marketing team. The overhead charges of using central facilities are also calculated based on product lives.
- *Is cost an issue with your products or are they one-of-a-kind?*: Any project considers its costs an issue tends to focus on monitoring and controlling its expenses more than a company that produces one-of-a-kind products, of which customers are willing to pay for any price. More than 57% of all companies thought their cost issues vary by products – mixture of unique and ordinary products. 36% of surveyed companies focused on costs of products whereas one-of-a-kind products makers accounted for only 7%.

In undertaking a new product development effort, a crucial question arises for the product with the realistic manufacturing cost targets; and that is, can it be sold at a profit?

At its simplest, this is merely a question of whether the price at which the product can be sold exceeds the cost at which it can be developed. The issue is really deeper than this because there is significant uncertainty about both price and cost. Further, the bias on the part of product advocates is to underestimate development cost and overestimate selling price.

In understanding cost there are three elements that must be considered: development expenses, continuing costs, and the lost profits of replaced products.

Development expenses include those that are fixed, for example, depreciation or laboratory equipment. These also are both certain and uncertain. Continuing costs are those associated with manufacturing the product when it is being produced and sold, including those associated with marketing, sales, and distribution. The likelihood of being financially successful at new product development is greatly increased by thorough and frequent estimation of all future costs. Financial analysis is relatively cheap compared to a failure.

If the proposed new product is a replacement for an existing product, one could have chosen not to undertake the new product development effort by continuing to sell the existing product. Although the sales and associated profit of the existing product might decline, this course of action would not require any unusual expense or cost on one's part. Thus, if a new product does make an existing product obsolete, one element of cost in new product development effort is the lost profit of the replaced product, which must be subtracted from projected profit of the new product.

In thinking about price, there really are two questions – One, is there a real market? And two what is the competition? The answers to these questions will establish the maximum price of the new product. To decide whether there is a real market, one must ask whether a need or want exists. If it does not, there must be a plan to create a need. But there are also many existing needs, the easiest to satisfy being a lower-priced substitute for a product that already is being purchased. After the existence of a need is established or the plan to create it is developed, it is necessary to ask whether customers really will buy the product. Do they have the money, and can the product be made available to them? Do they have any incentive to do so, or does it save them money?

The other issue that sets price is the existence of competition. If we are proposing to produce a product that could put another company out of business, that other company may react very aggressively to your entry into the marketplace. They probably are not just going to stand by and be put out of business. On the other hand, if there is not much competition, it is much easier to maintain a high price.

It is important to make sure that a new product development program being proposed in a company has a prospective financial return commensurate with the inherent risk it entails.

Although different companies use different financial justification techniques, many of them use the internal rate of return (IRR) to evaluate major investments. Thus, if new product development programs can be described in terms of IRR, these programs and their approvals are put on the same footing as capital investment approvals, for instance, whether to build a new manufacturing facility. When presented this way, new product development programs may show very attractive returns. But if these programs do not show attractive returns, perhaps they should not be funded.

The Profit MAP proposed by Rosenau (1982) has been used by business organizations for many years to calculate IRR, as also other financial measures. The Profit MAP (Measurements to Assess Programs) provides a format for rapidly analyzing a new product development program's profit potential. It enables identifying any critical areas requiring more detailed examination. As it summarizes the involvement of all departments in a company, one can use the Profit MAP to facilitate communication with the other functional managers involved in the NPD program.

9.1 Internal Rate of Return

IRR is one of many financial measures companies use to analyze the attractiveness of prospective investments. Profit margin, payback period, return on investment, and present value are some other financial measures, but each of these has drawbacks. Profit margin is simple but does not contain any information about the investment required to obtain the profit margin, nor does it provide any measure of risk. Payback period ignores everything that occurs after payback, though it is simple and useful indicator of risk because a long payback period is more risky than a short payback period. Return on investment (ROI) is an important measure, however, for future investments it can be done only on a period-by-period basis (for example, fiscal year) or it must be averaged over some longer time period. In common with the two preceding measures, ROI fails to account for the time value of money. Net present value (NPV), which does reflect the time value of money, is related to IRR, but NPV requires that a discount rate be pre-selected. Both NPV and IRR are forms of discounted cash flows (DCFs), in which money in future years is taken to have less value than today's money. Each of these techniques is treated in financial textbooks in more detail.

IRR depends on the timing of expenses and receipts, thus recognizing the time value of money. Calculation of IRR with the Profit MAP provides all the information to permit quick determination of the other measures. The inherent assumption in IRR is that recovered funding is reinvested into the same venture, as the venture's rate of return. IRR is the rate by which the cash flow must be discounted, such that the total discounted cash flow is equal to zero over the period representing the life of the investment, for instance, ten years. Expressed another way, IRR is the discount rate for which the program's NPV is zero. Thus, IRR provides a direct comparison against the cost of money and is the prevailing method employed in many financially sophisticated companies. Many business pocket calculators have an IRR function which can be used to calculate IRR.

To calculate IRR, we need to have the total cash flow for each year of the NPD program. To determine cash flow, we have to determine five estimated quantities for each year: (i) company sales resulting from the NPD program; manufacturing costs; development expense; operating costs; and capital expenditures. It is prudent to make these estimates before engaging in a NPD program rather than persisting in an unprofitable program. If we lack the information to make these five estimates, it is a clear danger signal that the proposed program's ultimate success depends on unknowns. The work required to make the estimates is an essential element of a successful NPD program. Because each of the five quantities is an estimate, the derived IRR is an estimate. However, it does not take that long to analyze a proposed NPD program's IRR and its sensitivity to the estimates. In doing this analysis, one gains a great deal of insight. The significance of IRR value is that the IRR value of alternative investments can be compared, and resources can then be committed to the most promising alternative. Any company's basic strategy, and the reasons for undertaking new product development should establish minimum IRR target for the company. As a general rule, forecasted new product development IRR values should exceed 20 percent, or perhaps 30 percent, because of the risk. If the forecasted IRR value is lower, there probably is a better investment available elsewhere.

There frequently are intangible factors not amenable to analysis. Therefore, one should use IRR as a guide rather than an absolute criterion. If a thoroughly analyzed new product program has a high IRR (say, greater than 20 percent), it is probably a good undertaking, even though it should still be judged against other investment opportunities. If the calculated IRR is low (say, less than 10 percent), look around for a better new product investment opportunity, or, if one goes ahead, understand the financial risk being undertaken. In between these extremes, one must recheck assumptions and not just change the numbers. In these intermediate cases, one may find it helpful to use the probabilistic techniques which can be found in financial textbooks. Do not, however, go ahead with a program that has a low IRR just because someone in R&D or marketing says new product development is inherently risky. Finally, if enough information is not available to estimate the IRR, that is a danger signal warning of that we do not yet know enough about the proposed NPD program.

9.2 Market Turbulence:

This section of the survey is anticipated to discover how market conditions have changed during the past decade and how the surveyed companies could adjust their NPD strategies and tactics to meet such changes. The changes of market demands and consumer behaviors toward diversity force manufacturers to adjust their new product strategies to meet the diversified requirements, at the same time, they have to keep product costs at a low level as well. One of solutions to cope with such market turbulence with cost effectiveness is mass customization technique that has been discussed in details in chapter 2. Based on the summary results in table 4-16, following are topics asked in the survey material and their responses.

- *To what extent are product demand levels unstable and unpredictable?*: The question was intended to lead into product demand levels, which are the first type of customer demands that change over time. The results showed that the demands (in volume) today are more fluctuating than those in the past decade.
- *What rate are the needs and wants of customers changing*?: At present days, customer demands are changing more quickly than they were in last 10 years. However, not every industry behaved in the same way; from survey results, companies in hard drive, instrument, and printing and publishing industries are facing fewer demand changes than they were in history.
- ***Prices** of products influence in customer decisions to buy*: On average, customers today seem to be more sensitive on prices of products than those in last 10 years. However, prices cannot influence customers of one-of-a-kind products much, and it looks like the prices of this type of products have less impact than before.
- ***Quality** of products influence in customers decisions to buy*: Not surprisingly, the product quality demands today is still highly influencing buying decision although they are not much different from those in last 10 years. The quality issue has been and is still ranking the first for buying decision.
- ***Fashion and style** influence in customer decisions to buy*: In general, this factor was ranked at a very low level in this survey. Only products in automobile and construction, that customers concern fashion and style, still maintained the importance. From the survey, the today's customers on average valued fashion and style more than that in the past.

Table 4-15. Summary of responses market turbulence issues by industry

Industry		Automobile	Biotech & Drugs	Computer Storage	Construct	Defense	Electronics & Electrical	Foods	Industrial Machinery	Instrument	Medical Devices	Printing & Publishing	Overall Average
Demand Fluctuation Level	- Last Decade	50%	0%	48%	15%	24%	37%	48%	39%	48%	44%	18%	38%
	- Today	50%	5%	44%	45%	48%	51%	60%	56%	64%	42%	20%	48%
Need Changing Rate	- Last Decade	20%	25%	51%	50%	36%	26%	25%	43%	50%	43%	37%	38%
	- Today	80%	5%	40%	85%	53%	46%	62%	63%	42%	54%	13%	50%
PRICE Impact to Buying Decision	- Last Decade	80%	5%	30%	70%	50%	52%	73%	60%	45%	39%	42%	48%
	- Today	90%	5%	65%	70%	36%	68%	68%	68%	47%	63%	28%	58%
QUALITY Impact to Buying Decision	- Last Decade	20%	5%	43%	80%	56%	70%	48%	58%	69%	86%	48%	62%
	- Today	70%	5%	46%	80%	59%	74%	47%	63%	62%	86%	43%	65%
FASHION Impact to Buying Decision	- Last Decade	20%	0%	18%	75%	5%	11%	15%	34%	5%	37%	28%	22%
	- Today	55%	5%	6%	85%	11%	15%	30%	43%	4%	39%	48%	26%
SERVICE Impact to Buying Decision	- Last Decade	20%	35%	61%	30%	43%	53%	30%	61%	61%	56%	65%	53%
	- Today	40%	5%	40%	90%	60%	54%	50%	67%	60%	59%	60%	56%
FLEXIBILITY Impact to Buying Decision	- Last Decade	20%	35%	40%	40%	30%	33%	33%	58%	48%	44%	50%	45%
	- Today	40%	15%	36%	90%	44%	39%	43%	62%	54%	58%	20%	49%
UNIQUENESS Impact to Buying Decision	- Last Decade	30%	25%	43%	50%	45%	34%	38%	48%	50%	43%	50%	46%
	- Today	55%	15%	29%	90%	41%	41%	45%	52%	53%	55%	13%	47%
INNOVATION Impact to Buying Decision	- Last Decade	30%	25%	50%	50%	53%	50%	35%	53%	45%	39%	28%	50%
	- Today	55%	5%	59%	50%	54%	64%	37%	67%	64%	56%	28%	55%
Customer dictate price, condition, features	- Last Decade	60%	5%	36%	40%	70%	57%	20%	52%	69%	86%	28%	51%
	- Today	40%	5%	60%	80%	70%	67%	42%	64%	66%	68%	43%	61%
Battle in Market	- Last Decade	60%	25%	30%	20%	46%	58%	43%	48%	5%	37%	58%	51%
	- Today	90%	5%	66%	85%	56%	76%	87%	74%	32%	91%	48%	70%
Competition on PRICE vs DIFFERENT	- Last Decade	30%	25%	35%	90%	41%	48%	35%	49%	61%	56%	48%	50%
	- Today	15%	15%	30%	35%	43%	48%	47%	43%	56%	59%	60%	46%
Product Life	- Today	3	5	0.81	3.5	25.13	6.25	5.5	6.58	11.2	4.28	6.13	7.53
	- Last Decade	5	7	2.38	35	24.25	10.08	12.75	7.75	13.25	6.5	11.17	10.51
Amount of Product Variety		10%	-30%	5%	45%	5%	26%	47%	30%	15%	19%	12%	19%
Average Product Life Cycle		20%	-50%	13%	30%	5%	14%	22%	5%	-3%	15%	0%	9%
Product Quality Level		30%	-45%	13%	10%	23%	21%	32%	27%	11%	18%	-17%	16%

Table 4-16: Summary of responses market turbulence issues by industry (Continue)

Industry	Automobile	Biotech & Drugs	Computer Storage	Construct	Defense	Electronics & Electrical	Foods	Industrial Machinery	Instrument	Medical Devices	Printing & Publishing	Overall Average
Product Costs	20%	-45%	-5%	15%	-15%	22%	17%	11%	3%	3%	-10%	**5%**
Customer Satisfaction	15%	-50%	9%	35%	20%	28%	18%	23%	10%	15%	-12%	**15%**
How Much Products are Customized	-40%	-35%	-8%	50%	40%	6%	2%	18%	21%	-1%	3%	**8%**
Product Customization Level	-5%	-45%	-3%	50%	2%	15%	8%	13%	4%	7%	12%	**8%**
Production: One-of-a-kind or Mass	50%	-25%	8%	-40%	-6%	9%	33%	16%	-25%	29%	-20%	**7%**
Process Flexibility Level	10%	-45%	0%	45%	5%	19%	23%	8%	14%	12%	10%	**11%**
Shorter or Longer NPD	20%	-	-10%	5%	18%	18%	23%	9%	10%	-5%	-3%	**7%**
PD Project Length - Today	2	12	13	24	75	14.33	12	10.5	20.4	24	5.25	**21.69**
PD Project Length - Last Decade	5	12	28.67	60	99	17.17	17.5	18.17	27	23.25	15	**30.54**
Tools to enhance product customization - IT	0%	0%	50%	100%	75%	38%	33%	83%	20%	33%	100%	**49%**
- Automation	0%	100%	50%	100%	25%	63%	67%	83%	0%	33%	67%	**49%**
- Lean Manufacturing	100%	0%	25%	100%	25%	38%	33%	33%	20%	56%	67%	**40%**
- Collaborative Eng	0%	0%	0%	0%	25%	0%	0%	17%	0%	0%	33%	**7%**
- NVAR	0%	0%	0%	100%	25%	13%	33%	33%	20%	33%	0%	**22%**
- SCM	0%	100%	25%	100%	0%	25%	67%	67%	60%	33%	0%	**38%**
- Modular Manufacturing	0%	0%	25%	0%	25%	38%	33%	50%	40%	22%	0%	**29%**

- ***Level of service** influence in customer decisions to buy*: In aggregate, service became a bit more important for buying decision nowadays than last 10 years. However, some industries deemed on the opposite way; especially biotechnology and hard drive industries considered their services were by far less than the past.
- ***Flexibility** of products influence in customer decisions to buy*: In general view, customers in last 10 years did not value product flexibility as much as today's customers do. Product flexibility can be referred to options and features that clients can choose to have in their products. However, companies in biotechnology and printing industries felt that they offered less flexible products than they did in the last decade.
- ***Uniqueness** of products influence in customer decisions to buy*: Product uniqueness can be defined as distinctiveness of products that is hardly imitated. This survey revealed that this factor was not different between customer perceptions during past 10 years.
- ***Innovation** of products influence in customer decisions to buy*: On average, customers today use innovation to support buying decisions more than people did in last 10 years.
- *Customers dictate the prices, conditions, and features of products*: In general, customers have more power to control product prices and features today than last 10 years. The major force that drives the change could be that customers now have more choices while producers are competing each other to gain more market share. Nevertheless, one-of-a-kind product makers still benefit from their products that can dictate their prices; for instance, customized recreational vehicles (RV) and medical devices.
- *You and competitors battle for market share in the markets*: The competition situation gets worse at the present time than 10 years ago. The oversupplied products plus merging and acquisition trends during the past decade boosted the competing battle to hyperactive markets.
- *Competition is based on product differentiation or on price competition*: The study found out that overall surveyed companies used price competition strategy little over product differentiation strategy at the present time, as seen at the total values in figure 4-21. Back to 10 years ago, both strategies were equally used. Construction was the industry that has shifted the most from extreme product differentiation in last decade to price competition nowadays.

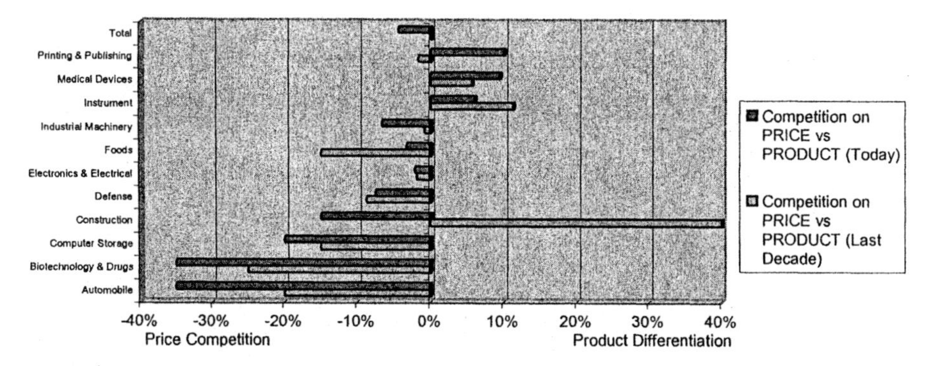

Figure 4-21. Comparison of competition strategies of industries during past 10 years

- *Product variation today compare with that of 10 years ago*: On average, surveyed companies had more product variation at the moment than 10 years ago by almost 40%. However, a company in biotechnology industry was exceptional.

Table 4-17. Comparison of product life cycles between today and 10 years ago

Industry	Product Life (Today)	Product Life (Last 10 yr)	% Change
Automobile	3	5	-40%
Biotechnology & Drugs	5	7	-29%
Computer Storage	0.81	2.38	-66%
Construction	3.5	35	-90%
Defense	25.13	24.25	4%
Electronics & Electrical	6.25	10.08	-38%
Foods	5.5	12.75	-57%
Industrial Machinery	6.58	7.75	-15%
Instrument	11.2	13.25	-15%
Medical Devices	4.28	6.50	-34%
Printing & Publishing	6.13	11.17	-45%
Overall Average	7.53	10.51	-28%

- *Average product life cycles today compare with that of 10 years ago*: Almost all firms participating in the survey thought that their products these days have shorter lives than those 10 years ago. The degree of difference can be on average 28% as shown in table 4-17. The longest product life in the surveyed companies was products for military that had not much product life change during the past 10 years. The most volatile product was computer-storage products with product life less than 1 year.

- *Product quality today compare with that of 10 years ago*: In aggregate, the quality level of products at the present time is higher than that of 10 years ago in a degree of 32%. Yet, companies in biotechnology and printing industry found that their products today have worse quality than last 10 years.
- *Production costs of your products today compare with those of 10 years ago*: There was not a clear consensus out of this question. Companies in biotechnology, computer storage, defense, and printing and publishing claimed their production costs raised compared to 10 years ago, while others had them lower. Anyway, the overall average costs have brought down by 10% over the decade.
- *Today's products compare with those of 10 years ago in meeting customer needs*: Customer satisfaction level of today's products was uncovered to be higher on average for all industries except biotechnology and printing and publishing. The improvement during the past decade accounted for 30% increment based on the survey data.
- *Level of products being customized to individual customers*: The overall results showed that products of the surveyed companies have been customized to meet individual requirements in a progressive rate. In other words, products in the surveyed industries have more variety today than 10 years ago except automobile, biotechnology, and computer storage industries.
- *Production process is one-of-a-kind production or mass production*: The survey results demonstrated the production process could be divided into two groups according to two types of one-of-a-kind products and mass products. However, within the same category, the participating companies had different levels of unique or mass production; for example, automobile industry fully used mass production technique whereas construction industry fully customized their products.
- *Production flexibility today compare with that of 10 years ago*: Once considering process flexibility to customize products, companies on average viewed that processes today are more flexible than 10 years ago at the level of 22%.

Table 4-18. Comparison of new product project length between today and 10 years ago

Industry	Today's Project Length (months)	Last Decade Project Length (months))	% Change
Automobile	2	5	-60%
Biotechnology & Drugs	12	12	0%
Computer Storage	13	28.67	-55%
Construction	24	60	-60%
Defense	75	99	-24%
Electronics & Electrical	14.33	17.17	-17%
Foods	12	17.5	-31%
Industrial Machinery	10.5	18.17	-42%
Instrument	20.4	27	-24%
Medical Devices	24	23.25	3%
Printing & Publishing	5.25	15	-65%
Overall Average	21.69	30.54	-29%

- *Product development project length*: From table 4-18, one can see that the product development cycle time has been reduced substantially (almost 30%) during the past decade. Almost all industries except medical device have reduced the project time. The defense industry is still leading in terms of longest project time compared to the rest. Companies making customized products; such as recreational vehicles, windows, armaments, and prints – have significantly reduced NPD cycle time during the past decade.
- *Tools or techniques to enhance product variety and customization*: Based on the results in table 4-16, information technology (IT) and automation were the most widely used tools to enable product variety among the surveyed companies. Lean manufacturing and supply chain management (SCM) were not far behind. Modular manufacturing, non-value added reduction, and collaborative engineering techniques were also applied for this purpose.

Chapter 5

FINDINGS

This chapter presents the analyses of surveyed data with respect to the propositions in chapter 2. The variables and analytical methods and tools used are briefly discussed whereas the results will be deliberated extensively in this chapter. The flow of contents in this chapter follows three main themes described in chapter 2.

1. THEME 1: TYPOLOGY OF NEW PRODUCT DEVELOPMENT

Proposition 1: *Companies pursuing both innovative and customer-responsive strategies for NPD tend to be more successful in business than companies engaging in only either one approach.*

1.1 Variables

As an independent variable, the responses to question 1 of Product Development Strategy section represented the Customer-responsive Strategy variable; while those for question 3 of the same section served as the Innovative Strategy variable. Each of these two driving factors was presumed to have different influences on product success. Thus, the third variable, named Combined Strategies, was derived from the average of first two practice levels to represent application of both strategies.

On the other hand, the business success can be referred to by successful launching of products in markets. Some studies discovered that the NPD performance can be classified to product success and project success and that both types of successes are not really correlated (Dooley et al 2001; Griffin, Page 1996; Griffin 1997; Lewis 2001). The answers to questions in the Product Success section of the questionnaire corresponded to the business success in this context. The summary of selected data records for this study is shown in table 5-1.

1.2 Analytical methods and tools used

This hypothesis was aimed to find relationships of two product development strategies and product successes in different aspects; hence, linear regression analysis was used to verify the relationships. SPSS's linear regression function was first employed to determine whether the variance could arise from normal random distribution rather than meaningful association – significance-F in ANOVA table must not be greater than 0.05. The Microsoft Excel provides the X-Y scattering graph with equation and R^2 for each pair of correlating variables. The coefficient in the regression equation determines the growth rate of product success per unit of strategy employed. Generally, the higher the R^2 value, the tighter the data fit to the predicted equation. In other words, the higher R^2 value enables the model to predict more accurately.

Table 5-1. Summary of data records used in studying correlation in proposition 1 (* denotes variables used in study)

Record ID	Fit customer needs*	Most innovative features*	Combined Strategies*	Meet profitability	Capture market share	Generate revenue growth	Provide unique benefits	Be innovative in market	Product Success*
001	5	4	4.5	2	2	2	1	3	10
002	4	3	3.5	2	3	3	4	3	15
003	5	2	3.5	2	1	1	2	2	8
004	4	2	3	3	3	3	4	3	16
005	5	4	4.5	3	2	1	5	4	15
006	5	3	4	1	2	2	3	3	11
007	4	3	3.5	3	3	3	5	3	17
008	5	4	4.5	4	5	4	5	5	23
009	4	2	3	2	3	2	2	2	11
010	3	2	2.5	2	3	4	3	2	14
011	5	4	4.5	4	3	4	3	4	18
012	4	3	3.5	4	3	3	2	2	14
013	5	4	4.5	4	4	4	4	4	20
014	5	4	4.5	3	2	2	5	4	16
015	4	3	3.5	2	2	3	3	3	13
016	3	2	2.5	3	3	3	4	2	15
017	4	4	4	2	2	2	4	4	14
018	4	4	4	4	3	3	4	4	18
019	5	3	4	4	3	3	2	4	16
020	5	3	4	3	4	4	2	2	15
021	5	2	3.5	4	4	4	3	3	18
022	5	4	4.5	5	5	2	4	4	20
023	4	4	4	2	4	3	5	5	19
024	5	4	4.5	3	2	3	4	3	15
025	1	3	2	2	2	2	2	1	9
027	3	1	2	2	1	1	3	1	8
029	5	3	4	3	4	2	2	2	13
030	3	1	2	3	2	1	2	1	9
031	4	3	3.5	2	1	3	3	3	12
032	5	3	4	2	2	2	4	4	14
033	5	2	3.5	2	4	4	4	3	17
034	5	4	4.5	4	4	4	4	4	20
035	3	2	2.5	1	4	2	3	5	15
036	4	4	4	4	0	1	5	4	14
037	4	2	3	1	2	2	1	3	9
038	4	1	2.5	4	1	2	3	1	11
039	4	5	4.5	4	4	4	4	3	19
040	4	5	4.5	5	3	3	4	3	18
041	5	1	3	3	4	4	4	4	19
042	4	5	4.5	2	3	3	3	4	15
043	5	5	5	4	4	4	5	5	22
044	5	3	4	3	3	3	4	3	16
045	4	2	3	1	1	1	2	3	8

1.3 Results

After considering significance levels of all ANOVA (as shown in table 5-2) that all relationships did not have normal distributing residual, each pair of correlation was plotted in a graph with its regressed line and equation as of figure 5-1.

Table 5-2. Summary of linear regression analyses of variables associated with proposition 1

Independent Variable	Dependent Variable	R	R-Square	Significant F	Constant	Coefficient
Product success	Customer responsive	0.4284	0.1835	0.0042	2.5296	1.9469
Product success	Innovative strategy	0.5391	0.2907	0.0002	9.2343	1.8328
Product Success	Combined strategies	0.6072	0.3688	0.0000	4.1424	2.9169

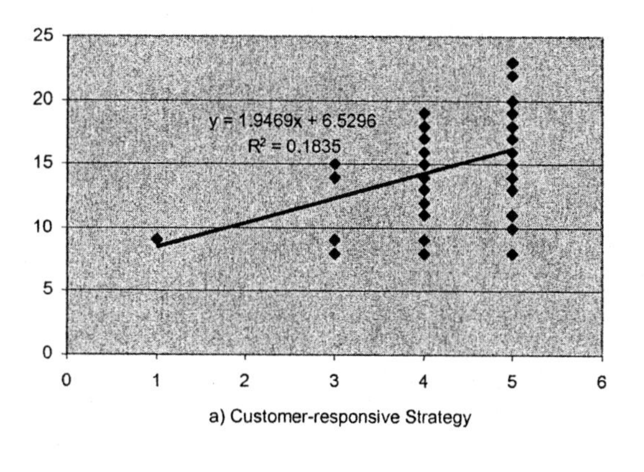

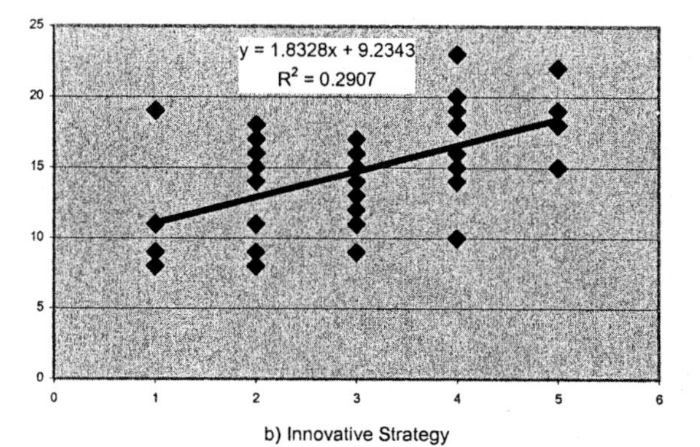

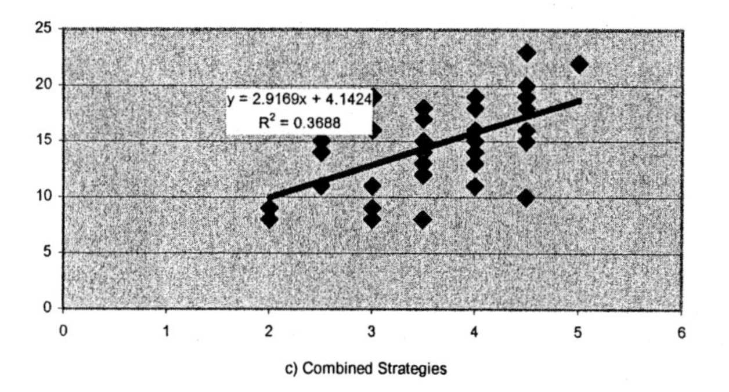

Figure 5-1. Linear regression analyses of NPD strategies on product success

Graph a) demonstrated that the relationship of customer responsive strategy and the summative product successes was not much correlated. Although the growth rate of product success based on customer-responsive strategy (slope of the line) was as high as 1.9469, the accuracy of the equation to stand for the actual data was too low ($R^2 = 0.1835$).

Graph b) introduced a model that better fit to data than the first strategy did ($R^2 = 0.2907$). The innovative strategy produced growth rate of 1.8328 times of success factor. Finally, the aggregate effect of two strategies could boost the product success factor at the rate of 2.9169 with $R^2 = 0.3688$.

From the surveyed data, it can be concluded that the combination of both customer-responsive and innovative strategies has positive impact to product success greater than employing each strategy alone. Therefore, the hypothesis was justified.

2. THEME 2: ROLES OF ORGANIZATION LEARNING AND KNOWLEDGE MANAGEMENT IN NPD

Proposition 2a: *NPD process success positively relies on implementation of organization learning and knowledge management.*

2.1 Variables

The independent variables used in this study were derived from responses of 7 questions regarding organization learning in the NPD Processes section and knowledge management in the Human Resource Development section. The dependent variable relating to NPD process success was the ability to reduce product development time that conformed to the responses of question 24 in the Product Turbulence section. The survey results were recalculated to percent of NPD cycle time reduction to be used in this study.

2.2 Analytical methods and tools used

At first attempt, all raw data were put into SPSS's linear regression analysis to study the relationship of individual independent variable and the process success factor. At that moment, the sample size reduced to 40 samples (as seen in table 5-3) due to insufficient data from some respondents. All the correlations were not acceptable since they all had high significance-F values.

Table 5-3. Summary of data records used in studying correlation in proposition 2a (* denotes variables used in study)

Record ID	NPD Process Training	Continuous Learning	Assess Training Needs	Training Gap Analysis	Monitor Training Status	Minimum Training Set	Company Pays for Trainings	Average OL+KM*	NPD Cycle Time (Today)	NPD Cycle Time (10yr)	% PDTR*
002	1	4	1	1	2	1	4	**2.45**	12	12	**0%**
003	1	3	1	1	1	1	3	**1.70**	24	15	**-60%**
004	5	4	3	3	3	4	5	**3.55**	30	72	**58%**
005	5	1	0	0	5	5	5	**2.25**	6	1	**-500%**
006	0	4	0	2	2	1	5	**2.40**	0.5	1	**50%**
007	2	3	2	3	2	1	5	**2.55**	14	24	**42%**
008	3	3	3	2	4	3	5	**3.20**	18	9	**-100%**
009	4	3	3	4	5	4	5	**3.70**	108	84	**-29%**
010	1	3	1	2	3	3	5	**2.70**	48	60	**20%**
011	1	2	2	2	2	2	5	**2.30**	8	12	**33%**
012	3	3	2	2	4	2	5	**2.75**	48	72	**33%**
013	1	1	1	0	1	4	4	**1.50**	6	12	**50%**
014	0	5	0	0	0	0	2	**1.45**	2	4	**50%**
015	3	5	3	4	3	3	5	**3.90**	60	120	**50%**
017	3	5	4	4	4	5	5	**4.45**	84	120	**30%**
019	1	2	0	1	1	2	1	**1.00**	12	18	**33%**
021	4	4	4	3	3	2	5	**3.50**	12	24	**50%**
022	0	2	1	2	2	1	5	**1.85**	24	60	**60%**
023	2	3	3	2	2	2	3	**2.70**	6	12	**50%**
024	3	1	2	0	0	0	5	**1.25**	24	24	**0%**
025	1	2	0	1	3	1	4	**1.60**	6	12	**50%**
026	2	2	2	2	3	3	5	**2.70**	8	48	**83%**
027	4	1	1	2	2	0	3	**1.50**	12	36	**67%**
028	2	2	4	2	1	1	3	**2.50**	7	12	**42%**
029	2	3	3	3	4	5	4	**3.50**	12	18	**33%**
031	4	2	2	1	0	2	2	**1.60**	36	18	**-100%**
032	0	3	1	1	0	0	4	**1.50**	8	12	**33%**
033	0	0	0	1	3	0	2	**0.70**	36	24	**-50%**
034	2	3	1	3	2	1	4	**2.20**	6	12	**50%**
035	5	2	2	4	2	2	5	**2.50**	18	24	**25%**
036	4	4	4	1	5	1	5	**3.50**	24	60	**60%**
037	4	4	4	1	4	2	5	**3.50**	18	24	**25%**
038	4	4	3	3	4	2	4	**3.45**	24	12	**-100%**
026	2	2	2	2	3	3	5	**2.70**	8	48	**83%**
038	4	4	3	3	4	2	4	**3.45**	24	12	**-100%**
039	5	5	5	1	3	5	5	**4.30**	30	15	**-100%**
040	4	3	3	4	3	2	5	**3.20**	18	23	**22%**
041	4	4	2	2	2	3	4	**2.80**	6	10	**40%**
042	4	3	2	1	2	2	3	**2.25**	6	12	**50%**
043	4	3	2	1	2	1	4	**2.35**	12	18	**33%**
044	2	3	0	0	1	0	4	**1.25**	18	24	**25%**
045	4	3	3	1	1	0	4	**2.40**	18	24	**25%**

After the first attempt, some data records with negative values of cycle time reduction were removed to better interpret the performance of majority group. Mostly in medical device industry, the isolated samples had product development time increased during the past decade. Other factors besides organization learning and knowledge management could really play a big part in the NPD cycle time characteristics of this group. The seven independent variables were amalgamated into one single factor, called 'Average OL+KM'. The remaining 32 sample records were normalized in order to reduce the scattering effects. The records were sorted by the degree of Average OL+KM and then separated into four groups of 8 records. At that moment, the average percent cycle time reduction was ready to be studied comparing to the ascending 'Average OL+KM'. The data records after normalization that looked different from table 5-3 are summarized in table 5-4.

Table 5-4. Summary of data records after normalization used for studying proposition 2a (* denotes variables used in study)

Record ID	NPD Process Training	Continuous Learning	Assess Training Needs	Training Gap Analysis	Monitor Training Status	Minimum Training Set	Company Pays for Trainings	Average OL+KM*	NPD Cycle Time (Today)	NPD Cycle Time (10yr)	% PDTR*
019	1	2	0	1	1	2	1	1.25	12	18	33%
032	0	3	1	1	0	0	4	1.35	8	12	33%
014	0	5	0	0	0	0	2	1.45	2	4	50%
013	1	1	1	0	1	4	4	1.50	6	12	50%
022	0	2	1	2	2	1	5	1.60	24	60	60%
025	1	2	0	1	3	1	4	1.65	6	12	50%
024	3	1	2	0	0	0	5	1.70	24	24	0%
044	2	3	0	0	1	0	4	1.75	18	24	25%
								1.53			**38%**

Record ID	NPD Process Training	Continuous Learning	Assess Training Needs	Training Gap Analysis	Monitor Training Status	Minimum Training Set	Company Pays for Trainings	Average OL+KM*	NPD Cycle Time (Today)	NPD Cycle Time (10yr)	% PDTR*
006	0	4	0	2	2	1	5	2.00	0.5	1	50%
011	1	2	2	2	2	2	5	2.05	8	12	33%
027	4	1	1	2	2	0	3	2.05	12	36	67%
028	2	2	4	2	1	1	3	2.10	7	12	42%
002	1	4	1	1	2	1	4	2.15	12	12	0%
034	2	3	1	3	2	1	4	2.35	6	12	50%
010	1	3	1	2	3	3	5	2.40	48	60	20%
023	2	3	3	2	2	2	3	2.45	6	12	50%
								2.19			**39%**
026	2	2	2	2	3	3	5	2.50	8	48	83%
007	2	3	2	3	2	1	5	2.55	14	24	42%
045	4	3	3	1	1	0	4	2.65	18	24	25%
043	4	3	2	1	2	1	4	2.75	12	18	33%
042	4	3	2	1	2	2	3	2.75	6	12	50%
012	3	3	2	2	4	2	5	3.00	48	72	33%
029	2	3	3	3	4	5	4	3.15	12	18	33%
035	5	2	2	4	2	2	5	3.25	18	24	25%
								2.83			**41%**
041	4	4	2	2	2	3	4	3.30	6	10	40%
040	4	3	3	4	3	2	5	3.45	18	23	22%
037	4	4	4	1	4	2	5	3.60	18	24	25%
036	4	4	4	1	5	1	5	3.60	24	60	60%
021	4	4	4	3	3	2	5	3.70	12	24	50%
015	3	5	3	4	3	3	5	3.80	60	120	50%
004	5	4	3	3	3	4	5	4.05	30	72	58%
017	3	5	4	4	4	5	5	4.20	84	120	30%
								3.71			**42%**

2.3 Results

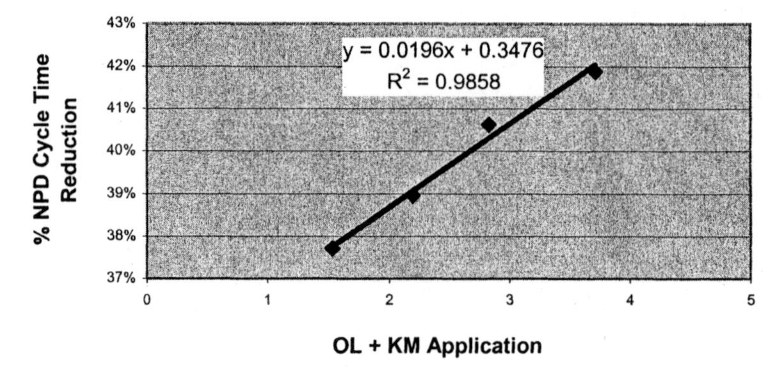

Figure 5-2. Linear regression analysis of organization learning and knowledge management on NPD process success

After a series of deliberate data analyses, the correlation of organization learning and knowledge management can be illustrated as in figure 5-2. After the outliers (all negative cycle time reduction data points) removal and normalization of data records, the regulated data were again put into regression analysis. The outcome at that moment was highly accurate with $R^2 = 0.9858$. The regression equation can be interpreted as every 10% increment of organization learning and knowledge management efforts contributes to 1% of cycle time reduction. Therefore, the hypothesis was justified with exception.

Proposition 2b: *NPD project success positively relies on implementation of organization learning and knowledge management.*

2.4 Variables

The independent variables used in this study were the same set as those of preposition 2a. The Average OL+KM value for each record was also computed. In this proposition analysis there were no data records deleted from the database because there were no missing data points. The dependent variable representing NPD project success was the on time ranking, which survey respondents had scored their average ability to complete projects on time. This variable can be used directly without any modification.

2.5 Analytical methods and tools used

Since the regression analysis of individual data records was not acceptable due to high variation, the data needed to be normalized before getting statistically investigated. The data records were sorted according to the average OL+KM values and then arranged into five groups of 9 data records. The average values of both independent and dependent variables of each group were put into a regression analysis as depicted in table 5-5.

2.6 Results

The outcome as seen in figure 5-3 was fairly acceptable since the R^2 was just 0.6091- not much accurate but not too bad. The equation can be interpreted that every 10% increment in learning effort can improve NPD project timeliness by 6.2%.

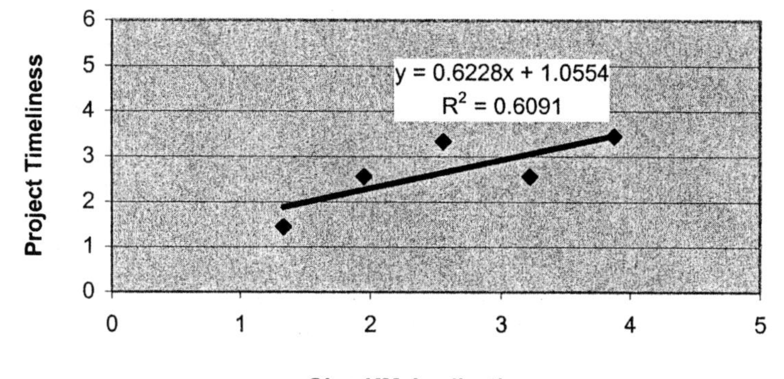

Figure 5-3. Result from linear regression analysis of organization learning and knowledge management on NPD project success

Table 5-5. Summary of data records after normalization used for studying proposition 2b (* denotes variables used in study)

Record ID	NPD Process Training	Continuous Learning	Assess Training Needs	Training Gap Analysis	Monitor Training Status	Minimum Training Set	Company Pays for Trainings	Average OL+KM*	On time Project*
033	0	0	0	1	3	0	2	0.86	1
014	0	5	0	0	0	0	2	1.00	1
019	1	2	0	1	1	2	1	1.14	1
032	0	3	1	1	0	0	4	1.29	2
030	2	0	1	0	0	2	5	1.43	2
044	2	3	0	0	1	0	4	1.43	2
024	3	1	2	0	0	0	5	1.57	0
003	1	3	1	1	1	1	3	1.57	2
013	1	1	1	0	1	4	4	1.71	2
								1.33	**1.44**
022	0	2	1	2	2	1	5	1.86	1
025	1	2	0	1	3	1	4	1.71	2
031	4	2	2	1	0	2	2	1.86	2
018	5	3	1	0	0	0	4	1.86	3
027	4	1	1	2	2	0	3	1.86	2
002	1	4	1	1	2	1	4	2.00	3
006	0	4	0	2	2	1	5	2.00	3
011	1	2	2	2	2	2	5	2.29	3
028	2	2	4	2	1	1	3	2.14	4
								1.95	**2.56**
034	2	3	1	3	2	1	4	2.29	4
045	4	3	3	1	1	0	4	2.29	2
042	4	3	2	1	2	2	3	2.43	4
007	2	3	2	3	2	1	5	2.57	3
043	4	3	2	1	2	1	4	2.43	3
023	2	3	3	2	2	2	3	2.43	4
010	1	3	1	2	3	3	5	2.57	4
012	3	3	2	2	4	2	5	3.00	3
016	5	3	3	1	2	2	5	3.00	3
								2.56	**3.33**
041	4	4	2	2	2	3	4	3.00	3
026	2	2	2	2	3	3	5	2.71	4
035	5	2	2	4	2	2	5	3.14	2
005	5	1	0	0	5	5	5	3.00	2
008	3	3	3	2	4	3	5	3.29	2
036	4	4	4	1	5	1	5	3.43	1
021	4	4	4	3	3	2	5	3.57	3
037	4	4	4	1	4	2	5	3.43	3
040	4	3	3	4	3	2	5	3.43	3
								3.22	**2.56**
029	2	3	3	3	4	5	4	3.43	2
038	4	4	3	3	4	2	4	3.43	3
004	5	4	3	3	3	4	5	3.86	4
015	3	5	3	4	3	3	5	3.71	4
001	5	4	5	4	2	2	5	3.86	3
009	4	3	3	4	5	4	5	4.00	4
039	5	5	5	1	3	5	5	4.14	4
020	4	3	2	5	5	5	5	4.14	2
017	3	5	4	4	4	5	5	4.29	5
								3.87	**3.44**

When this set of raw data were sorted by project on time ranking, the distribution of independent variable was very interesting as shown in figure 5-4. The diamonds shown in the chart symbolized mean values of degrees of learning implementation while the thick lines were acting for ranges of learning degrees. For example, at the category of project timeliness = 2, surveyed companies employed learning and knowledge management for 2.6 out of 5.0 scale (or 52%) within a range of 1.29 to 4.86 out of 5.0 scale. Considering the growing average OL+KM application levels along the ascending project timeliness can also assure acceptance of the hypothesis.

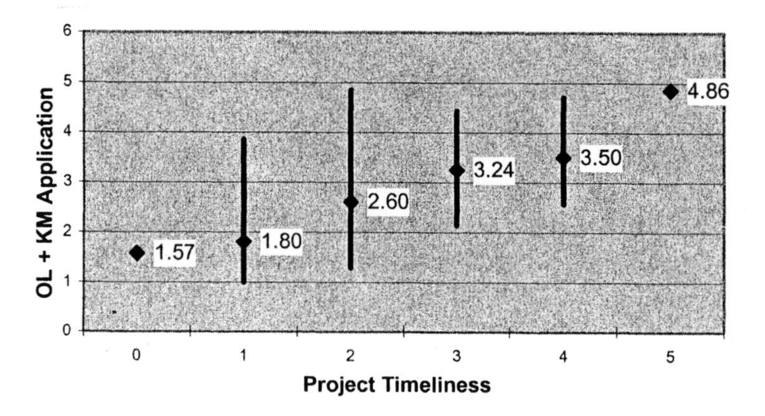

Figure 5-4. Means and ranges of OL+KM implementation based on project on time ranking

6. THEME 3: IMPACT OF MARKET CONDITIONS TO NPD STRATEGIES

Since the variables and methods used for analyzing correlations in this theme are the same for all 6 postulates, they will be described once while the result of each proposition is discussed separately.

3.1 Variables

There were two independent variable selected to study in this theme. First, percent product life reduction (%PLR) had been derived from the answers of questions regarding product life cycles today compared to last 10 years. Another independent variable was percent more demand fluctuation

(%MDF) that is computed from the ratio of the product demand fluctuation today minuses the demand fluctuation in 10 years ago over the fluctuation level in 10 years ago.

For the dependent variable, there were three factors proposed for study. First, the summation of product ratios in new-to-company and new-to-the-world categories served as the ratio of innovative products of a company. Second, % more product differentiation represented the marketing strategy that focuses more on product differentiation rather than on price competition. Last but not least, % product development time reduction was the last dependent variable studied in this theme. The detailed descriptions of these variables have been discussed in the chapter III.

3.2 Analytical methods and tools used

When raw data of all variables associated to this theme: - two independent variables and three dependent variables- were put together in SPSS regression analyses, they were all unacceptable due to too high significance-F values. Table 5-6 portrays the results of this initial study.

Table 5-6. Summary of linear regression analyses of variables associated with all propositions 3 (** Not needed if significance-F fails)

Independent Variable	Dependent Variable	R	R-square	Significance F	Constant	Coefficient
% PLR	% Innovative products	0.0545	0.0030	0.7382	**	**
% PLR	% More product differentiation	0.2574	0.0662	0.1297	**	**
% PLR	% PDTR	0.1674	0.0280	0.3085	**	**
% MDF	% Innovative products	0.0706	0.0050	0.6825	**	**
% MDF	% More product differentiation	0.1596	0.0255	0.3251	**	**
% MDF	% PDTR	0.0229	0.0005	0.8946	**	**

Consequently, all raw data in each hypothesis were sorted by ascending independent variable and normalized by grouping into five clusters before getting regressed. The mean values of each pair of analyzed variables represented raw data of the clusters. After that the regulated data were put into regression analyses. The results are as follows.

Proposition 3a: *Ratio of innovative products launched by a company positively relates to product life cycle reduction.*

3.3 Results

According to the SPSS regression analysis in table 5-7 and the Excel chart in figure 5-5, the data could not satisfy the minimum requirement for a good model in regression as evidenced by significance-F that was greater than 0.05. In addition, the coefficient obtained from the analysis, if credible enough, suggested a negative correlation of factors. Therefore, the hypothesis was not justified.

Table 5-7. Summary of linear regression analyses of variables in proposition 3a

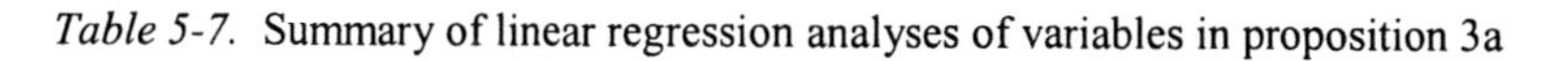

Independent Variable	Dependent Variable	R	R-square	Significance F	Constant	Coefficient
% PLR	% Innovative products	0.2863	0.2820	0.6405	30.976	-7.984

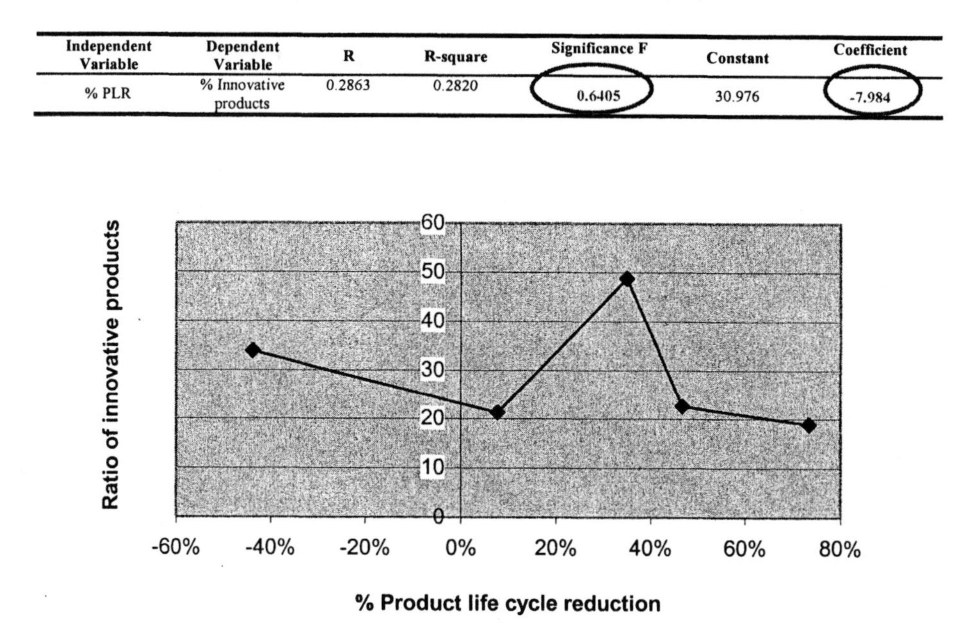

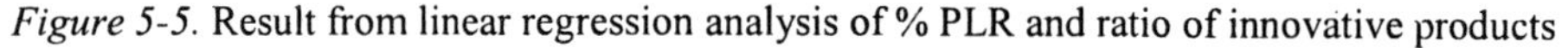

Figure 5-5. Result from linear regression analysis of % PLR and ratio of innovative products

Proposition 3b: *Companies base marketing strategy more on product differentiation rather than price competition when product life cycles shrink.*

3.4 Results

According to the SPSS regression analysis in table 5-8 and the Excel chart in figure 5-6, both factors were positively correlated. The model equation suggested that every 10% of product life reduction forced the companies to apply more product differentiation strategy by 2.92%. As a result, the hypothesis was justified.

Table 5-8. Summary of linear regression analyses of variables in proposition 3b

Independent Variable	Dependent Variable	R	R-square	Significance F	Constant	Coefficient
% PLR	% More product differentiation	0.9902	0.9805	0.0012	-0.0916	0.2921

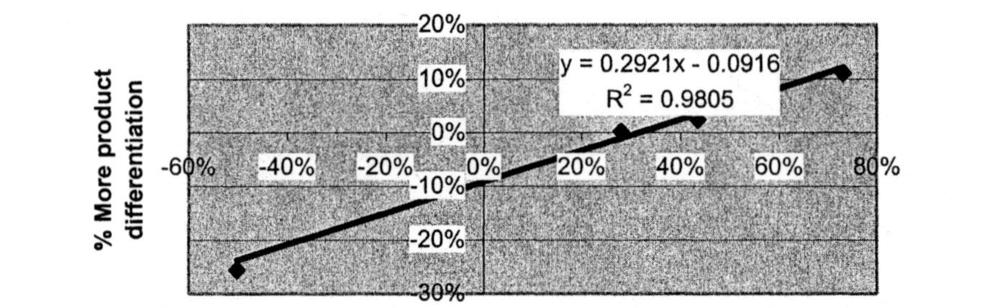

Figure 5-6. Result from linear regression analysis of % PLR and % more product differentiation strategy

Proposition 3c: *Product development times change in conjunction with product life cycles.*

3.5 Results

After normalization, the representative data points moderately fit into a straight line as shown in table 5-9 and figure 5-7. With a coefficient of 0.9139 in the model, it can be translated that product development times of surveyed companies were decreased at almost the same rate as the product life cycle reduction. In conclusion, the hypothesis was justified.

Table 5-9. Summary of linear regression analyses of variables in proposition 3c

Independent Variable	Dependent Variable	R	R-square	Significance F	Constant	Coefficient
% PLR	% PDTR	0.9693	0.9396	0.0064	-0.1523	0.9139

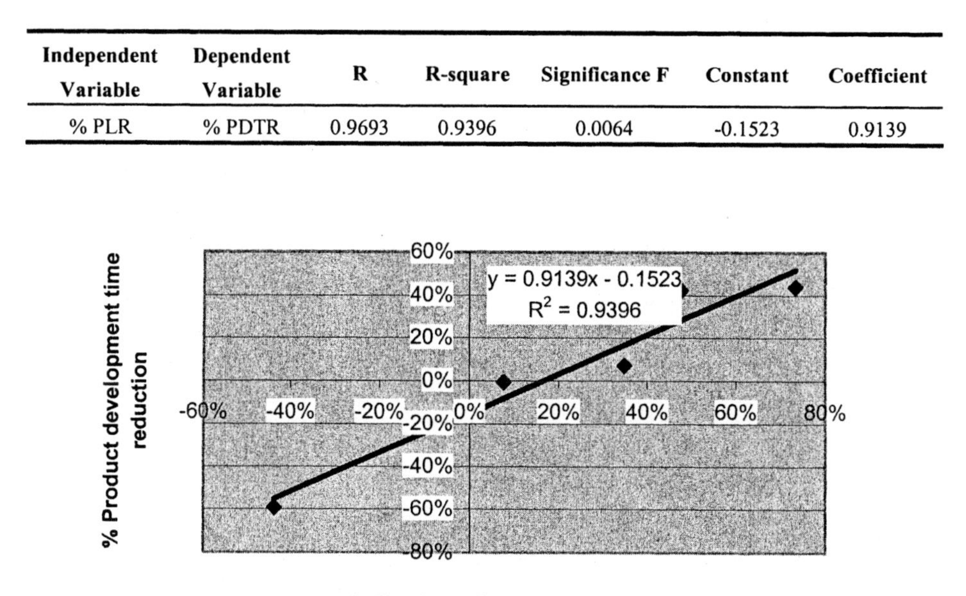

Figure 5-7. Result from linear regression analysis of % PLR and % PDTR

Proposition 3d: *Ratio of innovative products launched by a company positively relates to product demand fluctuation.*

3.6 Results

Even with normalization technique, the extremely high value of significance-F from ANOVA in table 5-10 can easily discard the hypothesis at the first glance. The figure 5-8 also confirmed that the data did not support the postulate well. It can be concluded that the hypothesis was not justified.

Table 5-10. Summary of linear regression analyses of variables in proposition 3d

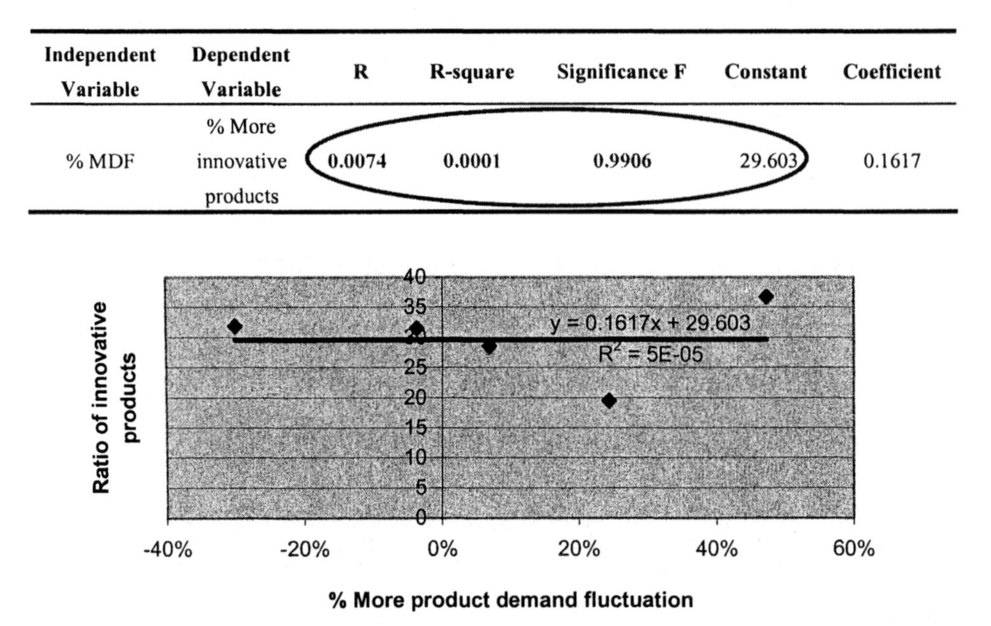

Independent Variable	Dependent Variable	R	R-square	Significance F	Constant	Coefficient
% MDF	% More innovative products	0.0074	0.0001	0.9906	29.603	0.1617

Figure 5-8. Result from linear regression analysis of % MDF and ratio of innovative products

Proposition 3e: *Companies base marketing strategy more on product differentiation rather than price competition when product demands are more fluctuating.*

3.7 Results

The significance-F value from ANOVA higher than 0.05 meant the distribution of residuals of real data points in relevant to the projected line was a normal distribution. Thus, the model failed to represent the studied data. Regardless of the low reliability of model ($R^2 = 0.5024$), the figure 5-9 shows that surveyed companies tend to implement product differentiation strategy rather than price competition when product demands are fluctuating. However, the hypothesis was not justified.

Table 5-11. Summary of linear regression analyses of variables in proposition 3e

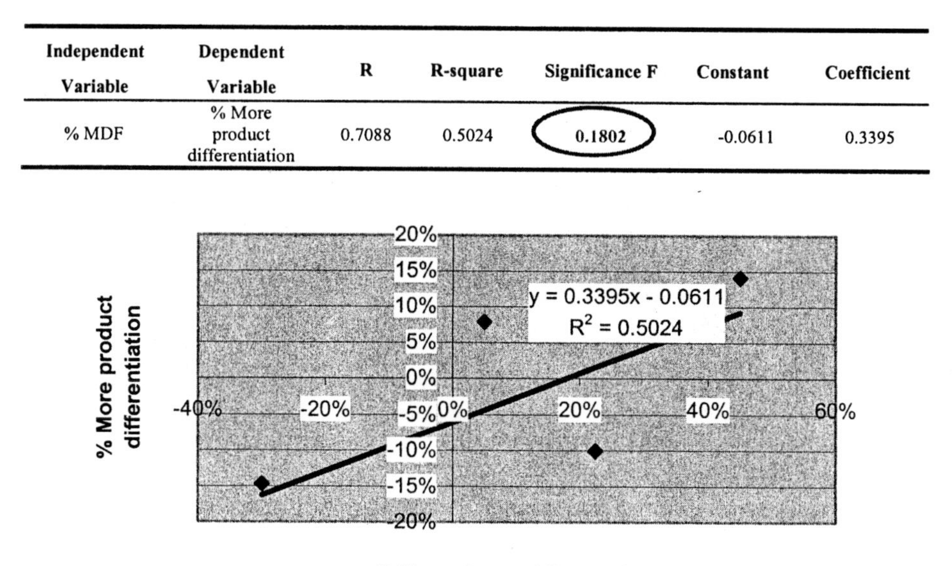

Independent Variable	Dependent Variable	R	R-square	Significance F	Constant	Coefficient
% MDF	% More product differentiation	0.7088	0.5024	0.1802	-0.0611	0.3395

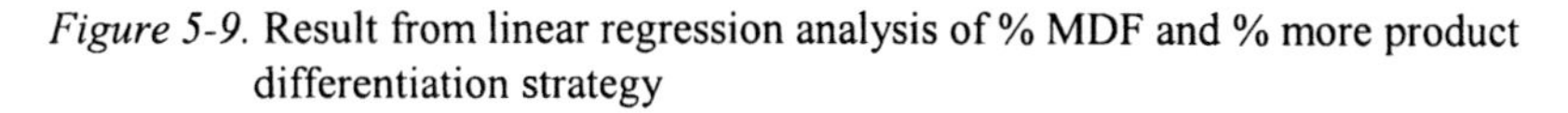

Figure 5-9. Result from linear regression analysis of % MDF and % more product differentiation strategy

Proposition 3f: *Product development times are reduced when product demands are more fluctuating.*

3.8 Results

The hypothesis was again rejected due to too high the significance-F value as seen in table 5-12

Table 5-12. Summary of linear regression analyses of variables in proposition 3f

Independent Variable	Dependent Variable	R	R-square	Significance F	Constant	Coefficient
% MDF	% PDTR	0.1204	0.0145	0.8471	-0.0182	0.1619

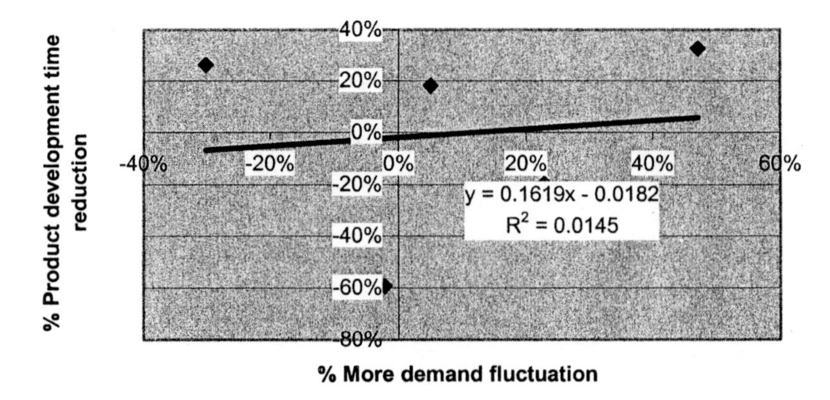

Figure 5-10. Result from linear regression analysis of % MDF and % PDTR

The overall conclusion of findings is discussed in the next chapter. The problems that prevented the study from being more accurate and suggestions for further studies are also discussed in the next chapter.

Chapter 6

CONCLUSIONS AND RAMIFICATIONS

1. CONCLUSIONS

A business unit that desires to survive and prosper in a hyperactive market environment today needs to be innovative as well as react exactly to what customers want. New product development (NPD) is crucial for business success since it is the process that determines quality, costs, prices, features, and market acceptance of new products launched into a market. There have been number of studies focusing on best practices and approaches aimed to enhance project-level performance [Cooper 1985; Dooley et al 2001; Flint 2002; Githens et al 2001; Griffin 1997; Griffin, Page 1996; Kuczmarski 2000; May 2001; Montoya et al 2000; Page 1993; Smith 1998; Smith 2000; Verworn 2000; Willaert et al 1998]. However, some of these best practices may be working well in some companies and some industries but not for all. Even in the same organization, these practices may only be used effectively for some projects. So, these best practices are not a real silver bullet. The root of problem is that the approaches are only valid in a small scope (project-level), but they are not applicable for a wider scope.

There are only a handful of researchers attempting to study in a wider context – to corporate-level strategies and implementation [Conceição et al 2001; Drucker 1998; Hamel 2003; Lewis 2001; Flower et al 2000; Senge 1998]. The scope of previous researches so far covers strategic competitiveness, innovation strategy, and learning organization. None of the strategic studies above really includes the new product development in its context. This research thus aims to study on how to implement corporate strategies in the NPD in order to achieve a sustainable competitive advantage.

From the study, there are two major strategies driving new product development: - innovation strategy and customer-responsive strategy. Companies implementing innovation strategy, known as *Prospectors*, are typically technology-driven and leading in new product introduction to market with new technology platform. Their competency takes root in the radical innovations that are hard for competitors to imitate and the ability to dictate prices and preferences of new products. On the other hand, companies following customer-responsive strategy, also known as *Reactors*, have different characteristics. They are basically market-oriented and are not the first-to-market but rather select the most wanted products and develop them in a more cost-effective way. Therefore, the reactors are more likely to have higher return on investment within a shorter time than prospectors do. However, to achieve both sustainable competency and also meet customer needs in the changing market environment today, a company should adapt the benefits of both strategies.

This study proposed several NPD-related hypotheses and used surveyed data from manufacturing companies for their validation. The findings from analyses of data in relation to proposed hypotheses are summarized as follows.

Proposition 1: Implementing both innovative strategy and customer-responsive strategy results in higher level of product success: - in terms of revenues, profit margin, market share, innovativeness, and customer satisfaction, than solely implementing either one of them.

Proposition 2: Both NPD process success: - measured from the decrease in product development time, and NPD project success: - identified by ability to complete projects on time, can be enhanced by improved organization learning and knowledge management.

Proposition 3: Among 6 sub-propositions introduced in the study; there were only two supported by the surveyed data. First is that the shrunk product life cycles in the current market situation influence the competing firms to have more product variety. Another validated sub-proposition is that the shorter product lives force the manufacturers to bring their products to the markets within shorter time. Authors' believe that the other four sub-propositions that failed due to insufficient data were:

- The product life cycle reduction should motivate companies to make more innovative products.
- The unstable level of product demands should force companies to develop more innovative products.
- The unstable level of product demands should drive companies to pursue more product differentiation rather than price competition.
- The unstable level of product demands should bring the product development time down.

2. LIMITATIONS AND RECOMMENDATIONS

In spite of high efforts to design research methodology and control the circumstances during the studying period, there are a number of obstacles that prevent the study from being perfect. In addition, anyone who wants to apply the results of this study should keep in mind about some conditions and limitations underlying beneath the results. The obstacles and limitations including recommendations are discussed as follows.

- *Too low survey response rate*: The study based partly on survey results, which was not well responded unless the respondents have a high interest in the new product development subject matter. There were only 22.1% of all invited persons responding to the survey. The major reason hindering the participation was the length of survey instrument even though most questions were designed to be easy to answer. Some respondents complained that the questionnaire was too lengthy and required a high degree of effort and thought to be filled out. Added into complexity, the survey questions covered a wide range of information that a normal employee of a company could hardly know all. Consequently, some respondents may have to ask someone else to exactly know the answers or just simply skipped the questions. Another possible reason could be there was no incentive for participating in the survey except being a recipient of a final report of the survey. We believe the results of this research study would be beneficial to those involved in product development including the respondents who participated in this study.

  To improve the response rate in the future study, the survey instrument should be carefully designed to be succinct and cover only one or a few aspects of NPD at a time. Applying information technology such as Internet-based survey will also facilitate the responses. The online survey can be designed to provide interactive results to the respondents immediately, which might motivate the respondents much more than promising that the results will be furnished to them three months later as being planned in this research study.

- *Geographical distribution of surveyed companies in the U.S. Midwest*: All survey participating firms are located in Minnesota and surrounding states in Midwest zone of the U.S. Someone might believe that the U.S. Midwestern have some typical characteristics, which might reflect in management and competition styles, different from other parts in the United States and other countries. But the fact is that these firms are not only local companies – some may involve only state wide, regional, national, or even multi-national markets. For that reason, these companies can represent companies all over the country.
- *Responses from manufacturing section with technical-bases*: All responses were obtained from technical-oriented professionals of manufacturing companies. This may cast a doubt whether the responses can act well for marketing and business units. Thus, anybody who wants to apply these study results must realize that the research was purposefully targeted to firms only in manufacturing sector.

To use the survey and analysis results effectively, one must realize the limitations and problems the research have encountered and select the most appropriate section to apply to his or her work. As mentioned before that the term 'Product' in this study includes manufactured goods and services, most respondents recognized the word as only tangible goods. Therefore, further researches on service innovations of manufacturing firms or new products of service industries; like financial, hospitality, and software development industries, would hold a great opportunity.

The following sections provide industry examples of NPD, how a technology roadmap is developed and also briefly reviews the collaborative NPD systems environment.

3. THE BUSINESS OF PRODUCT DEVELOPMENT

Two examples from computer (Hewlett Packard) and auto (BMW) industries describe the effective allocation of resources - people, process and technology - across the life cycle of a product.

3.1 Hewlett Packard (HP)

This industry application discusses customer-centered product design at HP. The company has established a Center of Competency in customer research and designing user interfaces, with facilities around the country. Its staff is educated and experienced in cognitive and physiological disciplines. The center provides centralized resources for HP design teams who do not have research, design and testing skills. It provides primary value addition to HP design teams through improved development process by bringing HP design teams together, creating a common product vision. It also enables bringing target users of products together with design teams to define, design and develop custom-centered products.

HP's Product Development Process focuses on user analysis, requirements and also product definition, design, and development for ease of use and usefulness. It also looks at other elements of the customer experience such as ordering, delivery, documentation, installation, integration with third party products and customer support.

For example, HP had a project to design a complex application suite, which required building storage network management. The challenges associated with storage network management include – developing very large scale, distributed networks with thousands of devices, huge amounts of data, very complex to manage, need continuous, reliable access to critical business data, and real-time monitoring and fast troubleshooting in the event of a hard drive crash. HP follows six steps product development – planning, requirements, design, development, system test and product release.

In the planning phase the challenges encountered were: vague product concept, customers' priorities were questionable, and developing a shared vision. Solutions to these challenges included forming customer focus groups, designed sketchy prototypes, customers filled in blanks. The values realized included understanding requirements, shared vision; not just a launch point for management applications; good idea of network representation and better view of business applications and data.

The challenges encountered in the requirements phase included: what details do users need and expect?; and requirements risks – incomplete, and failure to confirm. Solutions included – refined prototypes with more details and iterative research to define next level of detail. Values realized were details for information and task flow (for example, status, clicking behavior) and avoided costly re-design.

The challenges in the design phase were: complex – hundreds of technical details; and tradeoffs such as ease of use versus development cost. Solutions included: iterative user interface design, prototyping, testing; user interface specification, evaluation; cross-suite style guide. Value realized from this phase included continuous user focus keeps design usable and aligned with requirements; and cross-team coordination to ensure unified design.

The challenges in the iterative development phase were meeting somewhat unclear users' requirements and unanticipated design issues. Solutions included – user interviews to determine key features; numerous usability tests; structured expert review; and ongoing design consulting. The value derived included – customer-centered answers to design questions, and usability evaluation of actual product.

In the post-release phase, challenges included – establishing whether the product is easy to install and use or not; and whether it meets customers' real world needs. Solutions included – interviews with HP specialists in the field; and customer visits, and interviews. The value of usability data included – identifying gaps between user needs and actual product; and focusing on next version requirements on customer needs.

HP Corporation believes customer-centered focus can unify diverse stakeholders and be the basis for common process definition. They are also of the view that Customer-centered design activities can and should be incorporated into every phase of product design and development. Customer-centered design support activities are tailored to the needs of each project at HP. They have found that customer-centered design support activities decrease development time and cost by "getting it right the first time". Such activities increase customer satisfaction by meeting their needs and expectations.

3.2 BMW

This industry example relates to BMW auto group, a company known for their passion for innovation. Their first car was introduced in 1928. Their strong brand identity (apart from minor modifications since 1917) is the nucleus for strategic decisions and inward orientation includes BMW Group competencies and heritage. External expectations include brand values, product substance, and leading edge technology. BMW Group's Core Process Product Development links Innovation Management with the Car Development Process. Starting with the Strategy phase, the early phase is also called initial concept phase for the product, followed by Preparation phase prior to Car Series development. The Innovation Management process is a systematic, but flexible approach.

BMW group is a company with rich tradition founded in 1916. It sold more than 1,000,000 vehicles in 2002, and its more than 100,000 employees achieved a sale of about 42 billion Euros in 2002. Besides vehicles BMW also develops engines that have found their way up to the top of the Formula One racing cars. With the brand MINI, BMW Group also has successfully placed a quality automobile in the area of premium small cars. In addition, the brand responsibility for Rolls-Royce automobiles has been in the hands of BMW group since January 1, 2003. With the Rolls-Royce Phantom, BMW group continues a demanding tradition at the absolute pinnacle of the motor industry. Innovative ideas are one thing. A company's enterprise IT infrastructure forms the base for turning innovative ideas into reality. The use of the desktop and notebook computer varies widely. In addition to the typical office applications and some BMW specific applications, BMW's IT division must also meet the needs of designers who use workstations with high-end CAD software. BMW group's IT division faces the problem of finding a uniform, efficient and flexible computer architecture which can accommodate a wide range of user needs, offers stable images over a period of at least six months and supports the increasing mobility of employees in the future.

For its great innovation and sustainable growth by successful products, BMW group received the Outstanding Corporate Innovator (OCI) Award 2002 – the first European company to ever win – by the Product Development and Management Association (PDMA). BMW group has well-developed innovation management as well as its unique worldwide research and network with its integrated California Innovation Triangle (CIT). CIT includes the BMW Technology Office in Palo Alto, the BMW design studio Designworks/USA and the engineering and emission test center in Oxnard.

Under the leadership of the headquarters in Munich these research facilities combine new technologies together with high innovation character and customer use. Examples of the power of this technology network are the series introduction of the iDrive controller in the new BMW 7 or a new miniature head-up display for the Formula One helmet of Ralf Schumacher. BMW is a company on the go. Its determination to succeed is the result of a passion for mobility that is felt throughout the company, a continued drive for improvement. From this aspiration flows the energy that has made BMW group one of the world's leading automobile manufacturers.

BMW has a strong brand identity since 1917 and innovation is an obvious challenge for premium brands especially for BMW. Brand identity is the nucleus for strategic decisions in this company. BMW group's inward orientation provides competences and their heritage and external expectations drive brand values, product substance and leading edge technology. The controlling objectives for innovation management at BMW have been: Unique Selling Propositions (USPs) at launch of their models, breakthrough innovations each year, and concept cars each year. The company links innovation management with the car development process both during the strategy phase and the early phase of the development process. The Innovation Management Process follows a systematic but flexible approach. Various phases of innovation at BMW are: innovation research, innovation management and innovation transfer. Innovation Research is carried out employing Global Technology Scouting in US, Europe and Japan, Virtual Innovation Agency Process, which takes 6-8 weeks to come up with decision for recommendation.

Innovation management consists of forming innovation councils, evaluation and selection of innovative projects, allocation of resources, monitoring and reporting and it is an annual process of revisiting such projects selection. Innovation Council follows cross-functional approach and ensures that the model developed is based on concept cars and experimental vehicles, energy management, convenience and service, driving experience, lightweight, safety and security, and finally value and appeal. Members of Innovation Council include directors and senior managers from R&D, Group Marketing, Production, Corporate Affairs and Finance. Management of potential innovations is driven by prioritization in "must", "top" and "breakthrough" innovations categories. For each potential innovation and Innovation Sheet is prepared which provides technical project description, competition, market and customer assessment, strategic impact and financial situation. Innovation Management process is an annual cycle. It starts with a master list, concretization of innovation proposals, prioritization by must innovations, top innovations, and breakthrough innovations, planning of cost and manpower requirements, and finally decision taken on go/no-go focusing on customer benefit. The last phase of innovation management process is the innovation transfer. It consists of such steps as definition of vehicle USPs, identification of innovations ready for series development, and buy-in of vehicle projects. The first stage in this phase is the implementation of sound innovation concept. This includes: match to market requirements (USPs), project status with certificate of conformity (status, risk, planning, resources and economics), integration into complete vehicle, configuration compatibility, strategic objectives (Carbon Dioxide, weight, cost) and legal requirements. For managing risk, technological risk is balanced with market risk.

In summary, BMW group has a strategic commitment to innovation. It also has clearly defined innovation targets. The company employs cross-functional integration through the whole process and also has the fast access to the world's technology centers.

The following section outlines the application of Technology Function Deployment for developing Product Life Cycle plan and a technology roadmap.

4. TECHNOLOGY FUNCTION DEPLOYMENT, TFD

In a rapidly changing marketplace, enterprises must have a very clear understanding of their technology needs and opportunities. The enterprises' plans must be sufficiently detailed so that they identify what technology is needed, when the technology is expected to be used, and who will be involved in the efforts to develop it. Resources must be committed to assure that plans are met. The enterprise that expects to use technology as a potential tool to compete must assure that its technology plans are fully supportive of the overall business plan. The product cycle plan, which is part of overall business plan, shows detailed plans of new product launches, upgrades of existing products, and obsolete products which have completed their useful life. The final result of this planning process should be a roadmap of the enterprise's technology needs that involves a consensus understanding of the risks that are associated with this plan in terms of the timing, costs, and capabilities of the organization.

The approach to accomplish this roadmap will involve asking numerous questions concerning the needs for technology as well as exploring opportunities that technology may offer to the enterprise. To formalize answers to these questions, we use a number of matrices that encourage the exploration of the important questions by involving all segments of the enterprise. These matrices are tools to assist the enterprise and their principal value is not in their completion, but in the assistance they offer in assuring that questions are not omitted inadvertently. These matrices are part of Technology Function Deployment process, which is intended to provide managers of technology with a procedure that will help them to support the proper technology at the proper time with the proper level of resources.

The Technology Function Deployment process is implemented by constructing a set of four matrices and Technology Roadmap:

1. Matrix I, Technology Needs as Determined by Plans for Products, Processes, and Services
2. Matrix II, Consolidated Technological Needs Determined by Plans for Products, Processes and Services
3. Matrix III, Assessment of Opportunities Presented By Technology
4. Matrix IV, Need for New Engineering and Systems Tools
5. Technology Roadmap

Let us go over how you would complete various matrix templates. We provide here a real life industry example where the TFD process is applied. It pertains to enhancing cleaning capability of Advanced Technology Vacuum Cleaner. After each template is introduced the corresponding example matrix is illustrated for an Advanced Technology Vacuum Cleaner.

Matrix I, Technology Needs

The Technology Function Deployment Matrix I Template, used to specify the technology needs as determined by plans for products, processes, and services. Matrix I identifies and prioritizes technologies needed in support of the attribute to be incorporated in a given product.

We start by listing the following product cycle plan information:

1. Product Description
2. Expected Date of Introduction
3. Estimated Sales by year for first three years
4. Customer
5. Competitors
6. Competitor Product
7. Best-in-Class Product

We next use the following steps facilitate completion of matrix I:

- Consider an important attribute, which was introduced in this Product.
- List the associated developments needed to introduce this attribute in the proposed Product.
- Identify the supporting technology(ies) for each development initiative.
- Classify each technology: B for Base; CA for Competitive Advancement; and R for Revolutionary.
- Assign on a scale of 1 to 10 (1 – Low; 10 – High) the Availability of each technology.
- Assign on a scale of 1 to 10 (1- Low, 10 – High) the Capability of the firm to develop this technology.
- Assign on a scale of 1 to 10 (1 – Low, 10 – High) the Likelihood of success in utilizing such technology.

- Add the rankings of each technology relative to its Availability, Capability of the Firm to develop the technology, and its Likelihood of success in utilizing such technology.
- Provide preliminary assessment of technology. Should it be pursued aggressively (A) or Considered further (F) to evaluate or Delay (D) its implementation.

Table 6-1 illustrates a completed Matrix I for an Advanced Technology Vacuum Cleaner with the following characteristics:

Product Description: Advanced Technology Vacuum Cleaner
Expected Date of Introduction: January 1, 2003
Estimated Sales: Year 1 2003 15,000 Units
Year 2 2004 60,000 Units
Year 3 2005 75,000 Units
Customer: Residential/Commercial
Competitors: Hoover, Eureka
Competitor Product: None on Market
Best-in-Class Product: Hoover WindTunnel™ Bagless - U5750900

Table 6-1. Technology Function Deployment Matrix I Vacuum Example

New Attributes of the Proposed New or Modified Product, Process or Service	Developments Needed for New or Modified Product, Process or Service	Technology Requirements to Accomplish Developments Successfully	Classification of Technology	Ranking of Technology Low=1 to High=10				Preliminary Assessment of the Technology
				Availability	Capability	Probability of Success	Sum of Rankings	
Improved Cleaning	Improved Suction	Impeller Optimization	R	3	4	6	13	A
		High Efficiency Motor	CA	7	8	8	23	A
	High Efficiency, useful attachments	Design Effort	B	8	9	10	27	A
	Improved Filtration	More Efficient Filter	B	5	4	3	12	F
		Investigate New Filter Media	CA	3	5	3	11	D
	Brush Agitation	Beater Bar	B	9	10	10	29	A
		Automated Height Adjustment	R	2	3	5	10	F
		Brush Oscillation	CA	7	6	7	20	A

Classification of Technology		Preliminary Assessment of the Technology	
B	Base	A	pursue Aggressively
CA	Competitive Advancem	F	considered Further
R	Revolutionary	D	Delay

Matrix II, Consolidated Technological Needs

The Technology Function Deployment Matrix II Template is used to specify the Consolidated Technological Needs Determined by Plans for Products, Processes and Services. This matrix consolidates the technological needs consisting of various products (new or enhanced) planned for development and introduction to market. Worksheet template rows should be added to or removed from as required to define the product. Table 6-2 illustrates a completed Matrix II for an Advanced Technology Vacuum Cleaner.

Table 6-2. Technology Function Deployment Matrix II Vacuum Example

NEW ATTRIBUTES of the PROPOSED NEW or MODIFIED PRODUCT, PROCESS or SERVICE	DEVELOPMENTS NEEDED for NEW or MODIFIED PRODUCT, PROCESS or SERVICE	TECHNOLOGY REQUIREMENTS to ACCOMPLISH DEVELOPMENTS SUCCESSFULLY	Classification of Technology	Life Cycle of Technology	Ranking of Technology Low = 1 to High = 10							Date Technology is Needed	Final Action	Action Plan if yes, by whom?
					Current Status	Capability	Probability of Success	Value to the Enterprise	Adaptability to Several	Cost to the Competitors	Sum of Rankings			
Impeller Optimization	Material Change	Injection Molding	CA	T	3	4	6	7	8	6	34		F	
	Geometry		CA	T	2	3	4	6	8	10	33		A	
High Efficiency Motor	Low Friction Bearings	Teflon Seals	B	T	7	8	8	7	5	3	38		A	
	Variable Speed Motor	Sw. Reluctance Control	B	P	2	2	4	9	7	5	29		F	
Design Effort	Low Friction Surface	Microreplication	CA	P	4	5	6	8	7	9	39		F	
Add Beater Bar	Existing Market Assess.	None	B	M	9	10	10	5	2	3	39		F	
Brush Oscillation	Mechanical Design	3D-CAD Modeling	CA	T	8	8	9	8	9	4	46		A	
		Rapid Prototyping	CA	T	7	8	9	10	9	4	47		F	

Matrix III, Assessment of Opportunities

The Technology Function Deployment Matrix III Template is used to specify the Assessment of Opportunities Presented By Technology. This matrix provides assessment of opportunities presented by various technologies (other than those identified in matrices I and II), which are of interest to the enterprise.

Following steps facilitate completion of matrix III:

- List the technologies of interest to a given enterprise. Do not list technologies identified in Matrix I and Matrix II.
- Indicate which product attribute the identified technology will support.
- Indicate what level of knowledge (High, Medium, or Low) of the technology the firm has.
- Indicate what is the Probability of successful development of the technology (High, Medium or Low).
- Identify the level of maturity of the technology (P – primitive, T – in transition, and M – mature) that will be challenged by this new technology.
- List the key problems to be addressed prior to implementation of the new identified technology.

- List how you will have to respond to the competitor for the new technology (Urgent, Moderate, None).
- Indicate by Yes or No the desirability of the technology for further consideration.
- Indicate who will be responsible for the detailed assessment of technology.
- Indicate mode of exploration of technology (I for internal, E for external, or J for Joint).
- Indicate who will be responsible for the action plan.
- Indicate who will have final responsibility for the Program.
- Indicate the reviewing authority for the Program (may be some committee).

Table 6-3 illustrates a completed Matrix III for an Advanced Technology Vacuum Cleaner.

Table 6-3. Technology Function Deployment Matrix III Vacuum Example

Proposed Technological Opportunities	New or Modified Products, Processes, or Services that Technology May Allow	Level of Knowledge	Probability of Success	Life Cycle of Competing Technology	Key Problems that must be Solved Before Technology Can Be Used Successfully	Assessment of Competitors Response	Desirability Further Consideration	Detailed Assessment (If yes, by whom)	Mode of Exploration	Action Plan (If yes, by whom)	Final Responsibility for Program	Review Authority
Variable Speed, Higher Efficiency Motor	Better Suction, Lower Power Required, Easier Vacuuming of Loose Rugs	High	High	M	Cost, MTBF	Protect Existing Technology	High	Yes- Engineering Marketing	Internal	Yes- Engineering, Marketing	Electrical Engineering	V.P. Engineering
Low Friction Bearings	Better Suction, Longer Life	Medium	Medium	T	Cost	Protect Existing	Low	No		n/a	n/a	n/a
High Strength Plastic Impeller	Lighter Weight, Better Suction	Medium	Medium	T	Strength, Materials Aging	Investigate New Technologies	Medium	Yes- Engineering	Joint	Yes- Engineering	Mechanical Engineering	V.P. Engineering
Use of Recycled Plastics	Better for the Environment	Low	High	T	Availability, Strength, Materials Aging	Investigate New Technologies	Medium	Yes- Engineering, Marketing	External	Yes- Engineering	Mechanical Engineering	V.P. Marketing
Bio-Degradable Bag Material	Better for the Environment	Low	Low	P	Availability, Durability, Shelf Life	Protect Existing Technology	Low	No		n/a	n/a	n/a
Dirt Sensing Technology	Better User Knowledge of Cleanliness	High	High	M	Cost, MTBF	Protect - Technology Already Deployed	High	Yes- Engineering, Marketing	Internal	Yes- Engineering Marketing	Electrical Engineering	V.P. Engineering
Optimized Impeller Design	Better Suction	Low	Low	T	ROI, Amount of Improved Airflow	Protect Existing Technology	Low	No		n/a	n/a	n/a

Matrix IV, Need for New Engineering and Systems Tools

The Technology Function Deployment Matrix IV Template is used to specify the Need for New Engineering and Systems Tools. This matrix lists various technologies that are needed for the development of the identified engineering and system tools.

Following steps facilitate completion of matrix IV:

List the engineering and system needs for a given enterprise.

Indicate new tools needed to satisfy engineering and system needs of the enterprise.

- List the tasks required to provide the new tools for the enterprise.
- In the next three columns, a ranking on the basis of 1 to 10, with the highest being 10, is to be provided of three items that will strongly influence whether the enterprise will choose to given further consideration to the exploration of technology.

  -*Adequacy of the technology* applies to whether the current level of development of the technology is adequate for the new tool.

  -*Availability of technology* applies to whether someone who will refuse or will demand an unreasonable price for making it available controls, the technology or whether it can readily be made available for your use.

  -*Expected lifetime of the tool* refers to the probable period of time that the tool will be useful to the organization. A tool that can be expected to be used over a long period of time will have a high ranking.
- Add the rankings of each technology relative to its adequacy, availability, and expected lifetime. A high value suggests that the technology and the expected value of the tools over a long period of time are good.
- Next, a preliminary conclusion is to be made as to the desirability of proceeding further. This decision is based on perceived needs, the fact that new tools are appropriate, that the tasks are required to provide the tools are not excessive, and the sum of the rankings for the technology.
- Try to ascertain whether a detailed technology assessment is needed.
- Establish a realistic data for completion of the tool.
- We then decide on mode of exploration of the tool, which could be internal, external or joint based on such factors as the availability of the proper internal technical capability, an understanding of your system by a group outside the company, the desirability and capability of controlling the dissemination of results, and probable costs of each mode.
- Based on the previous information, a decision is needed about the desirability of pursuing the development of the tool or delaying further consideration.
- If it is the conclusion that further action is warranted, indicate the need for the plan that will provide preliminary details concerning timing and costs.
- Identify who will be responsible for carrying out the final program.
- Identify the person or committee who will have continual purview over the progress of the program, recognizing that the development of some tools requires a long-term commitment.

Table 6-4 illustrates a completed Matrix IV for the advanced Vacuum Cleaner.

Table 6-4. Technology Function Deployment Matrix IV Vacuum Example

Engineering and System Needs	New Tools Needed to Satisfy Needs	Tasks Required to Provide the New Tools	Ranking of Technology Low=1 to High=10				Desirability Further Consideration	Detailed Assessment (If yes, by whom)	Expected Date of Completion	Mode of Exploration	Recommended Action	Action Plan (If yes, by whom)	Final Responsibility for Program
			Adequacy	Availability	Expected Lifetime of Tool	Sum of Rankings							
Noise Modeling	Noise Simulation Software	Product Evaluation, Vendor Evaluation, Training and Procurement	8	10	8	26	High	No	10/1/2001	Joint	Consider Further	Yes- Mechanical Engineering	Mechanical Engineering
Air Flow Modeling	Air Flow Simulation Software	Product Evaluation, Vendor Evaluation, Training and Procurement	9	10	8	27	High	Yes- Mechanical Engineering	9/1/2001	Joint	Aggressive	Yes- Mechanical Engineering	Mechanical Engineering
Calibrated Particulate Measurement System	Particulant Measurement Apparatus	Product Evaluation, Vendor Evaluation, Training and Procurement	5	2	7	14	High	Yes- Mechanical Engineering	2/1/2002	Joint	Consider Further	Yes- Electrical Engineering	Electrical Engineering
Accelerated Plastics Life Testing	Oven, Chemical Bath	Define Tests, Evaluation Oven, Evaluation needed chemicals, Training and Procurement	10	10	7	27	High	Yes- Mechanical Engineering	1/1/2002	Internal	Aggressive	Yes- Mechanical Engineering	Mechanical Engineering
Plastics Durability Testing	Oven, Test Fixture	Define Tests, Training and Procurement	5	10	6	21	High	Yes- Mechanical Engineering	10/1/2001	Joint	Aggressive	Yes- Mechanical Engineering	Mechanical Engineering
Accelerated Motor Life Testing	Load Simulator	Define Tests, Evaluation System and Procurement	10	10	8	28	High	No	12/15/2001	Internal	Aggressive	No	Electrical Engineering
Noise Measurement	Calibrated Sound Measurement System	Competitive Evaluation, Procurement	10	10	10	30	High	No	11/1/2001	External	Aggressive	No	Mechanical Engineering

Technology Roadmap

The Technology Roadmap Template presents the technology plan or roadmap that the enterprise will follow in the years ahead. It provides a description of all technologies that will be under study by the enterprise in the time frame extending to at least the end of the cycle plan for the new products and perhaps longer for the exploration of new technological opportunities and for the development of new engineering tools technologies. This is what is illustrated in the following figure showing technology roadmap for the enterprise. It is at this point that final budgets and plans for each technology must be finalized.

Figure 6-1 illustrates a completed Technology Roadmap for an Advanced Technology Vacuum Cleaner. The Technology Roadmap indicates the time period during which each activity should be accomplished.

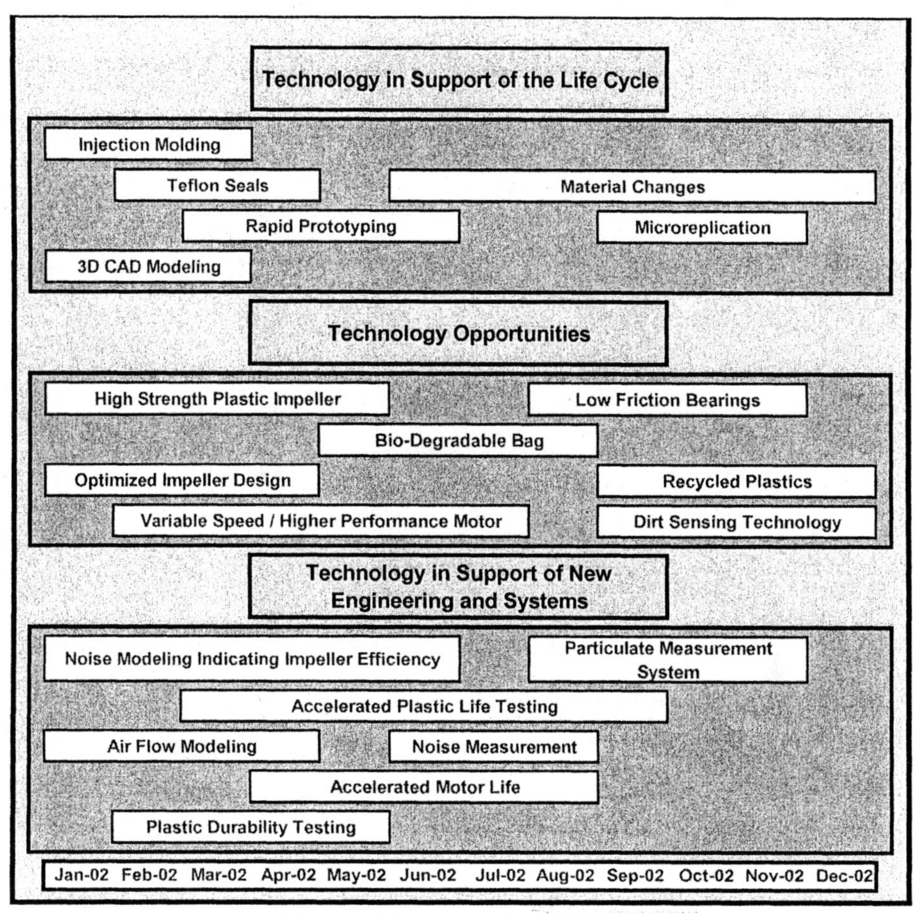

Figure 6-1. Technology Roadmap

We strongly believe TFD is an important tool for companies in coming up with the strategic product plan (for identifying and planning for technologies to support products and processes).

5. COLLABORATIVE NPD SYSTEMS

In the following section, common pitfalls and success factors observed in a collaborative NPD system environment are outlined. A brief discussion on what is considered to be an ideal NPD system is also presented.

5.1 Common Pitfalls/Success Factors

Based on authors' experiences, the following pitfalls are typical in collaborative NPD projects. These include: contentious supplier/customer relationships, unclear customer requirements, missing competencies, animosity (that is, not addressing natural fear of job loss), missing capacities and realizing the fact that leveraging knowledge requires intrinsic knowledge of what is to be leveraged.

Most significant success factors in collaborative NPD projects include: strong sponsorship from top management, strong project leadership, processes alignment and compatibility, systems compatibility, communication flexibility, and cultural sensitivity.

5.2 Ideal NPD Systems

The infrastructure of successful NPD system should include transparency with integrated Product Data Management/Product Lifecycle Management and Enterprise Resource Management systems; facilitation of knowledge management leading to re-use and sustainability; hierarchical project management; facilities for collaboration (such as, firewalls, etc.); full integration of financial elements; and hierarchical metric system with drill-down capability.

Such new product development efforts will demonstrate high level of performance. These systems are based on previous management concepts like PACE enhanced through application of web technologies and integrated across the enterprise by totally linking product creation from idea to market enabling teams across multiple geographies and cultures.

In conclusion, there is no single right way to do New Product Development. It needs to be looked at as a system. The focus is needed in the front and back end of New Product Development to enable speed, integration, collaboration and sustainability. Quantum leaps will occur with the true integration of existing technologies and tools into cohesive enterprise-wide systems.

Appendices

Appendix A

January 15, 2003

To Whom It May Concern:

As part of research study currently in progress, we are interested in gathering information on how on New Product Development approaches used by number of different manufacturers. The motivation for this study stems from the state of current market environment where is characterized by shorter product life cycle, fast-changing technology, globalization, and greater customer expectations.

The input of this study is mainly allocated to industrial surveys by means of collection of information on new product development from various companies in such industries as automobile, computer components, household appliances, electronics consumer goods, semiconductor, tools and machinery, and medical devices. The knowledge extracted from the survey will be compared with existing academic theories and shared among educational and industrial institutions.

The survey instrument, mostly in multiple-choice and ranking forms, is categorized in following 12 sections:

- General information (18)
- Product success (5)
- Project timeliness and schedules (20)
- Product development strategy (6)
- Customer interface (21)
- New product development process (23)
- New product research (8)
- Human resource development (6)
- Financial monitoring and control (15)
- Teamwork and leadership (10)
- Technology deployment (5)
- Market turbulence (25)

I anticipate the results of the survey will lead to a comprehensive report on New Product Development within next 3-4 months. The report will present the survey results in descriptive and statistical analysis in addition to a number of graphs and charts. Importantly, the report will be prepared in such a way that results from an individual company cannot be extrapolated against the competition in the same industry. The data of individual records will be kept strictly confidential.

I would appreciate if you could please have the survey completed by a person associated with new product development in your organization. The filled-out questions can be returned to my e-mail address at skumar@stthomas.edu.

Thank you in advance for your participation in this valuable survey project.

Sincerely yours,

Sameer Kumar, Ph.D. PE.

APPENDIX B: Questionnaire
Confidential New Product Development Processes Survey

Instruction: Please fill out this survey electronically and send via e-mail to skumar@stthomas.edu. All information you provide specific to your company will be held in confidence.Any questions on survey material, please contact Dr.Sameer Kumar at phone (651) 962-4350 or fax (651) 962-4710 for clarification.

General Background Information

This section collects general information of the surveyed companies and related industries and markets

Survey Date: ________

1. Name of Person Taking Survey ________________
2. Title ________________
3. Company Name ________________
4. Division of Company ________________
5. Federal Standard Industrial Code(s) for your Division or Company ____________
6. Number of employees in your company ________
7. Number of employees in your division ________
8. Is your company public or privately held? ________
9. What is the annual revenue of your company/division? ________ / ________
10. Number of employees involved in new product development________
11. Number of individuals with engineering title involved in new product development ________
12. Describe your industry in terms of product innovation. (Please check all that apply)
    ___ Fast changing ___ Slow changing ___ Stagnant
    ___ High competition ___ No competition ___ Many mergers & acquisitions
    ___ Legacy products ___ 1 product ___ 10 products ___ 100 products
13. Who are regulatory agencies relating to your industry and market?
    ___ FDA ___ EPA ___ FAA ___USDA ___ Others (Please specify) __________
14. Describe the average product life of products in your company/division ________
15. Do most of your product development cycles fall into the category of incremental change or platform change? ________________
16. Are your products primarily software, electronic, mechanical, chemical, other, mix? ____________
17. Some of new product ideas never reach commercialization. What percentage of your total new product ideas can reach each stage of the product development process below?

Idea Generation	Idea Screening	Business Analysis	Development	Test & Validation	Commercialization
100 %	%	%	%	%	%

Product Development Strategy

This section studies strategic factors pertaining to new product development in your company

Following are motives persuading firms to develop new products. Please consider your relevant product development strategy and rank the motives by filling in the ranking number in the blank.

0 – Never	1 – Few/Little	2 – Medium to Low	3 – Medium to High	4 – Mostly	5 – Always

Strategy	Ranking
1. To best fit customer needs	
2. To gain lowest product cost	
3. To have most innovative features	
4. To be First-to-market / First-to-the world	
5. To leverage company's revenue / profit / market share	

6. Referring to the definitions of new product development strategies, please roughly specify the percentage of each category based on all of your existing products.

Cost reduction ___ %
Market repositioning ___ %
Addition to existing lines ___ %
New-to-company ___ %
New-to-the-world ___ %
100 %

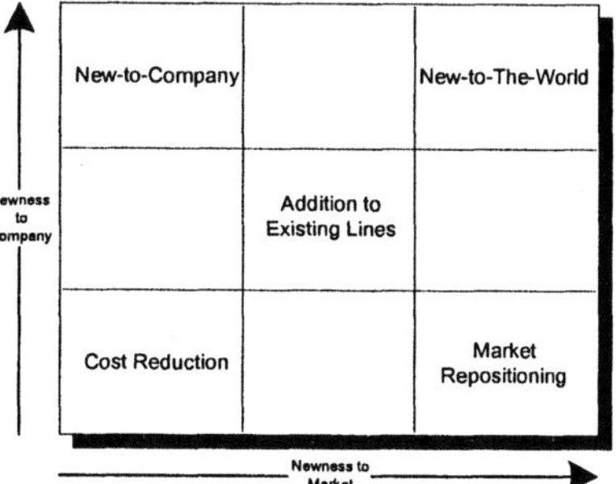

Product Success

This section collects information on your new product success to compare with your practices

Please rank following events based on actual results happening in your new product development

0 – Never	1 – Few/Little	2 – Medium to Low	3 – Medium to High	4 – Mostly	5 – Always

Event	Ranking
1. Your new products meet profitability targets	
2. Your new products capture significant market share	
3. Your new products generate significant revenue growth	
4. Your new products provide customers unique benefits	
5. Your new products are the innovations to market	

Customer Interface Issues

This section studies how customer requirements and feedbacks are used in your new product development process

Please rank following activities based on actual application in your company/division

0 – Never	1 – Few/Little	2 – Medium to Low	3 – Medium to High	4 – Mostly	5 – Always

	Activity	Ranking
1.	You have a formal process for evaluating customer needs prior to evaluating new product development opportunities	
2.	You use written, detailed product specification from customers to develop new product specification	
3.	You have a customer feedback method during pre-specification development phase	
4.	You have a customer feedback method during product development phase	
5.	You have a customer feedback method during prototyping or trial phase	
6.	You have customers directly involved in the development process	
7.	You let customers have access to detailed design information	
8.	You let customers have access to pre-production models or prototypes	
9.	You use customer-based alpha or beta testing processes to acquire feedback and performance testing information	
10.	Your project/program team have direct customer contact regarding new product development	
11.	You use techniques to simulate the customer environment during development	
12.	You involve customers in product development risk mitigation issues	
13.	Your customers generally know next innovation or technology that they will need to be competitive	
14.	You have to sell the customer on the next technology or innovation direction	
15.	Your customers have direct design database access	
16.	Customer requirements frequently change during the new product development cycle	
17.	When there is a requirement change, you inform all staff throughout the organization	

<u>Customer Interface Issues</u> (continue)

Please answer the following questions by selecting the relevant answer(s) from the list

Question	Choices
18. What product identification/specification tools do you use in product development process?	___ Quality function deployment (QFD) ___ Six-sigma ___ Fish-bone diagram ___ Taguchi method ___ Others (please specify) ___________
19. Who determines the minimum acceptable quality level of the new products in your industry?	___ Industrial standard ___ Company ___ Customer ___ Others (please specify) ___________
20. Are quality standards changing in your industry? Increasing? Decreasing?	___ Increasing ___ Decreasing ___ Stable ___ Vary upon products

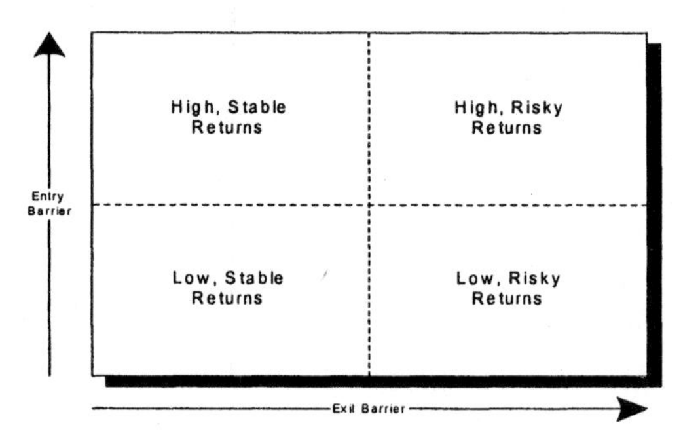

21. The diagram above represents a landscape of marketing and competition. Please consider the market and competition in your industry and select <u>only one</u> of those quadrants that most fits to your industry.

___ Low entry barrier, Low exit barrier => High competition, Easy to exit => Low, stable returns
___ High entry barrier, Low exit barrier => Low competition, Easy to exit => High, stable returns
___ High entry barrier, High exit barrier => Low competition, Hard to exit => High, risky returns
___ Low entry barrier, High exit barrier => High competition, Hard to exit => Low, risky returns

New Product Development Timeliness and Schedules

This section studies how you use project timeliness and schedules for new product development

Please rank following activities based on actual application in your company/division

0 – Never	1 – Few/Little	2 – Medium to Low	3 – Medium to High	4 – Mostly	5 – Always

Activity	Ranking
1. You use formal schedules and/or standardized baseline schedules for new product development	
2. Your company offers training on these tools either internally or externally for project team	
3. You apply project time compression technique in order to have formal and informal overlap of activities	
4. Your NPD schedules are detailed and broken down into the smallest possible work packages	
5. Your NPD team and management recognize that there is a linkage between PD and an effective supply chain	
6. Rate your organization's ability to complete projects / programs on schedule	
7. Customer needs influence the product development timeline	
8. The fuzzy front-end concept is recognized by management and personnel involved in product development	
9. Modular techniques and interface design between modules are considered in the early stages of the PD process	
10. Customers are given tradeoff choices in specification for shorter schedules? i.e. unit cost for quicker process	
11. Your NPD organization feedbacks schedule changes to its customers as well as the rest of the company	
12. Project schedules are used as a realistic tool for project / program tracking in your organization	
13. You are frequently asked to commit to your 'best case' schedule or an even tighter schedule	
14. Your team or personal performance is measured as a function of schedule performance	
15. You have initiatives in place to actively monitor and decrease product development cycle time	
16. You utilize collaborative (Internet based) engineering tools in their product development efforts	
17. You use baseline schedules for different types of projects	

New Product Development Timeliness and Schedules (continue)

Please answer the following questions by selecting the relevant answer(s) from the list

Question	Choices
18. What scheduling tools and methodologies does your company employ?	__ MS Project __ MS Excel __ Stage-gate Squeaky Wheel __ Other (please specify) ________
19. Are your customers typically given the 'best case' schedule, 'worst case' schedule, or both?	__ Best case __ Worst case __ Realistic case
20. At what point in a new product development cycle do you start measuring schedule performance?	__ Conceptualization __ Research __ Feasibility __ Design __ Prototyping

New Product Development Processes

This section studies how formal development processes and engineering tools are implemented in your new product development process

Please rank following activities based on actual application in your company/division

0 – Never	1 – Few/Little	2 – Medium to Low	3 – Medium to High	4 – Mostly	5 – Always

Activity	Ranking
1. You use written, well documented, procedures for new product development	
2. Your company provides training on these procedures	
3. Your significant suppliers and customers are involved in new product development from the start of the project	
4. You use (Internet-based) collaborative engineering techniques	
5. You use a modular approach to product design	
6. You use Failure Mode and Effect Analysis (FMEA) methodologies	
7. You use formal risk mitigation planning techniques in new product development	
8. You use formal disciplined design flow techniques	
9. There are program gates that need to be met to advance to the next phase of the project	
10. Your design processes are formally documented	
11. You have a formal system in place to diagnose and correct problems in the design process	
12. Continuous learning is incorporated into your project teams and company	
13. You use a PDM (product data management) system or similar systems	
14. Capacity planning and resource allocation are considered in the new product development process	

New Product Development Processes (continue)

Please answer the following questions by selecting the relevant answer(s) from the list

Question	Choices
15. How do you use a task force team to cross-check and/or improve the NPD projects?	___ Steering committee ___ Peer review ___ Phase review ___ Expert review ___ Other (please specify) ______
16. What accountability measures are used to assure product success (performance)?	___ Periodic review ___ CPM/PERT ___ Cpk ___ Financial performance ___ Other (please specify) ______
17. What metrics are used to measure performance?	___ Time ___ Quality ___ Cost ___ Efficiency ___ Other (please specify) ______
18. Are your design teams co-located or virtually-located?	___ Co-located ___ Virtually-located
19. If you use virtually-located teamwork, do all members of the product development team use the same design database?	___ Yes ___ No ___ Not applicable
20. How does your company keep its employees skills up to date with the industry?	___ Internal training ___ External training ___ Supplier training ___ Industry conference ___ Other (please specify) ______
21. What is the average size of an NPD project team including the cross-functional components	___ persons
22. If you are using a formal product development process, how long have you been using the process?	___ < 1 year ___ 1-5 years ___ 5-10 years ___ > 10 years
23. Describe tools you use for communication across locations or with customers and suppliers	___ Internet ___ Intranet ___ Satellite ___ Teleconference ___ Virtual Private Network (VPN) ___ MS NetMeeting ___ Others (Please specify) ______

New Product Research:

This section studies how research and development (R&D) activities support the product development process

Please rank following activities based on actual application in your company/division

0 – Never	1 – Few/Little	2 – Medium to Low	3 – Medium to High	4 – Mostly	5 – Always

Activity	Ranking
1. You use feasibility studies to determine the likelihood of success on technologies or in areas lacking of expertise	
2. You define a specific research phase on a NPD project	
3. You have a separate group of employees that perform most research activities	
4. You do most of the research activities during the new product development cycle	
5. You limit the amount of risk or unknowns that it takes on as part of a new product development opportunity	
6. Product innovations are part of the research or development phase in your company	

Please answer the following questions by selecting the relevant answer(s) from the list

Question	Choices
7. Do you use internal expertise or external expertise to research and develop new technologies?	__ Internal expertise __ External expertise
8. How important is it to manage market risks as opposed to technical risk in the NPD process?	__ Market risk is higher __ Technical risk is higher __ Both are important at the same level

Teamwork and Leadership

This section studies how NPD teams and leaders are selected and formed

Please rank following activities based on actual application in your company/division

0 – Never	1 – Few/Little	2 – Medium to Low	3 – Medium to High	4 – Mostly	5 – Always

	Activity	Ranking
1.	Your company has a formal project/program management role/title	
2.	Your company uses program mentoring	
3.	Your company use program sponsorship	
4.	Your company use cross-functional teams in product development	
5.	The executive staff actively involve in the direction or steering new product development activities	
6.	Engineers in your company are responsible and accountable for the outcome of the project	

Please answer the following questions by selecting the relevant answer(s) from the list

	Question	Choices
7.	Where do new product ideas come from?	___ Executive ___ Employee ___ Customers ___ Suppliers
8.	How are project teams and project leaders selected or teams formed?	___ Skill ___ Seniority ___ Availability ___ Other (please specify) ___ ________
9.	How are individuals and teams rewarded for meeting goals?	___ Recognition ___ Financial rewarding ___ Non-financial rewarding ___ Other (please specify) ___
10.	How many project/program managers do you have in your company/division?	___ persons

Human Resource Development:

This section studies how human resource development is arranged in your organization

Please rank following activities based on actual application in your company/division

0 – Never	1 – Few/Little	2 – Medium to Low	3 – Medium to High	4 – Mostly	5 – Always

Activity	Ranking
1. Your company formally assesses internal expertise and training needs of employees prior to launching NPD activities	
2. Your company has a formal training gap analysis process	
3. Your company actively monitors the training status and needs of each employee	
4. Your company has training or job readiness standards	
5. Your company has a minimum set of formal training that each new employee must complete prior to working on NPD projects	
6. Your company pays for seminars, training classes, tuitions for qualified employees	

Technology Deployment

This section determines how technology functions are deployed relatively to new product development in your organization

Please rank following activities based on actual application in your company/division

0 – Never	1 – Few/Little	2 – Medium to Low	3 – Medium to High	4 – Mostly	5 – Always

Activity	Ranking
1. You consider technology deployment is a key factor of an NPD success	
2. You develop a product cycle plan or a product roadmap	
3. Your products closely follow the product roadmap	
4. You develop a technology roadmap for your products, processes, and other systems i.e. IT, automation, etc.	
5. You deploy technology management: - technology forecasting, selection, transfer, and termination	

Cost / Profit Margin/ Return

This section studies how financial indices are monitored and controlled during new product development process

Please rank following activities based on actual application in your company/division

0 – Never 1 – Few/Little 2 – Medium to Low 3 – Medium to High 4 – Mostly 5 – Always

Activity	Ranking
1. You recalculate ROI, IRR, or market capitalization for a given project/program if the program schedule slips	
2. Projects/programs have ever been rejected or stopped if the financial return is not adequate	
3. You use a hurdle rate, min ROI or min IRR for determining whether or not to pursue an NPD opportunity	
4. Actual historical ROI or IRR is measured after a product is put into production in order to determine its success	
5. Your products are labor intensive to build	
6. Your company has a minimum acceptable profit margin on new products	
7. PRODUCT costs are actively monitored during the development process	
8. PROJECT costs are actively monitored during the development process	
9. Project capital costs (such as fixtures, prototypes, molds) are scheduled prior to new product development launch	
10. You continuously monitor capital costs, capital at risk, impaired capital, and capital usage	
11. You have ever delayed a product development project due to the inadequate cash flow to pay for capital items	
12. You monitor capital outlay to cash flow requirements	
13. You use Activity Based Costing and Analysis	
14. You compute and track product life cycle costs based on reliability data	

Please answer the following question by selecting the relevant answer(s) from the list

Question	Choices
15. Is cost an issue with your products or are they one-of-a-kind, demanding any price the end-users will pay?	___ Cost is the issue ___ Our products are one-of-a-kind ___ Cost issue varies by products ___ Other (please specify) ________

Market Turbulence

This section studies how the turbulence level in a market in the past decade could affect the new product development strategy

I. **Instruction**: Each of following questions asks for your perception of where your company/division stands *today* and *10 years ago* on a scale between two defined endpoints. Please answer by moving the mark for today situation and the mark for the situation 10 years ago by using mouse or just click on it and use cursor keys to move along the line to the point that matches your initial perception.

1. To what extent are the demand levels of your products unstable and unpredictable?

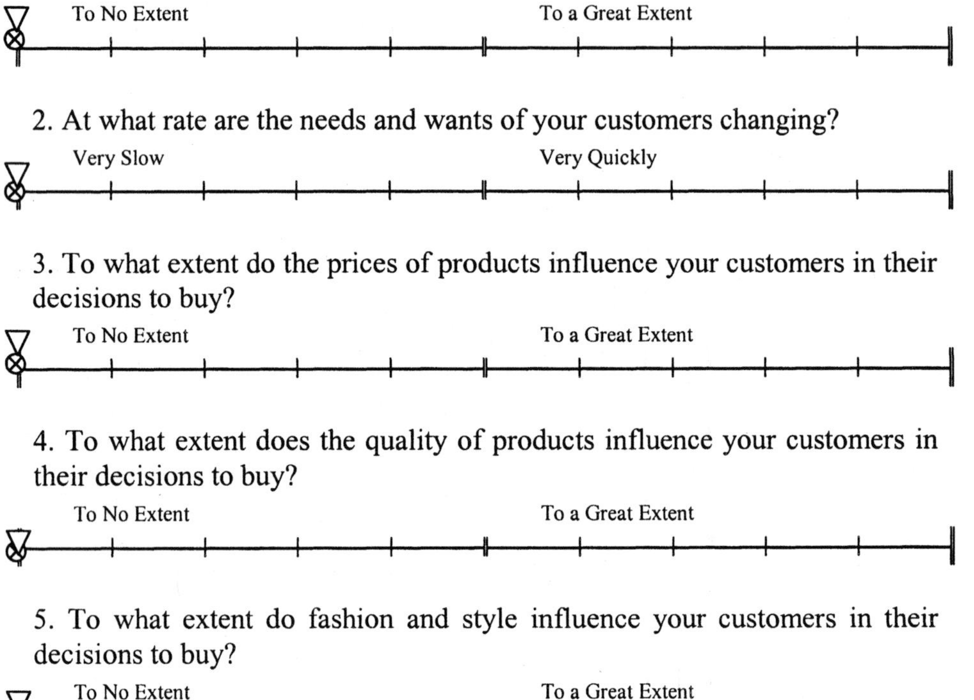

2. At what rate are the needs and wants of your customers changing?

Very Slow — Very Quickly

3. To what extent do the prices of products influence your customers in their decisions to buy?

To No Extent — To a Great Extent

4. To what extent does the quality of products influence your customers in their decisions to buy?

To No Extent — To a Great Extent

5. To what extent do fashion and style influence your customers in their decisions to buy?

To No Extent — To a Great Extent

6. To what extent does the level of service influence your customers in their decisions to buy?

To No Extent — To a Great Extent

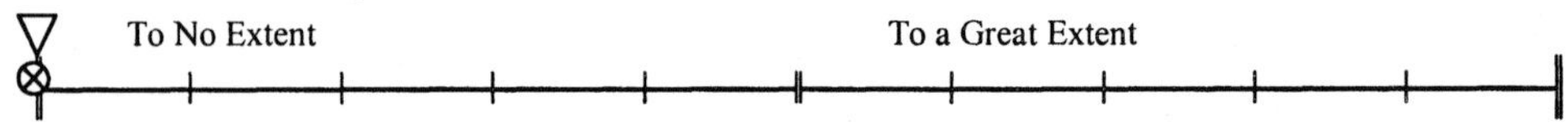

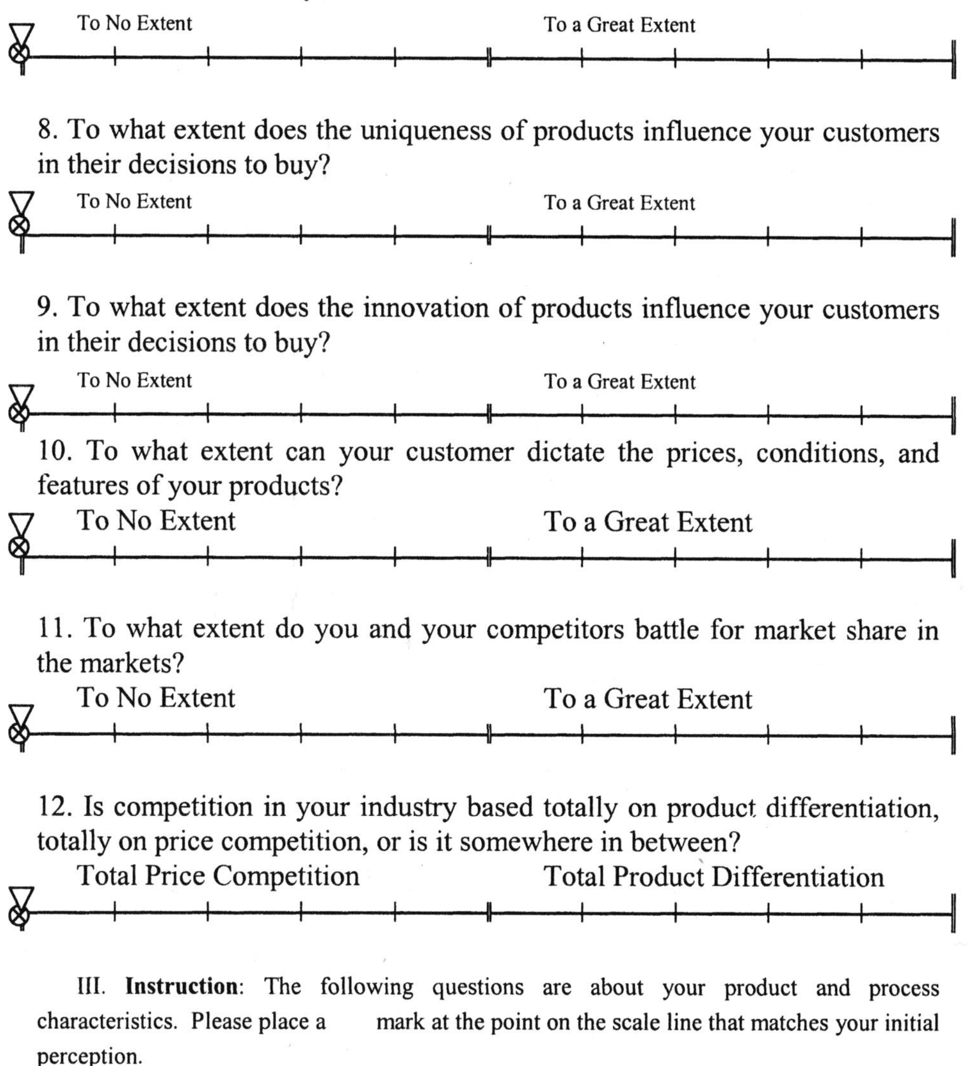

7. To what extent does the flexibility of products influence your customers in their decisions to buy?

To No Extent — To a Great Extent

8. To what extent does the uniqueness of products influence your customers in their decisions to buy?

To No Extent — To a Great Extent

9. To what extent does the innovation of products influence your customers in their decisions to buy?

To No Extent — To a Great Extent

10. To what extent can your customer dictate the prices, conditions, and features of your products?

To No Extent — To a Great Extent

11. To what extent do you and your competitors battle for market share in the markets?

To No Extent — To a Great Extent

12. Is competition in your industry based totally on product differentiation, totally on price competition, or is it somewhere in between?

Total Price Competition — Total Product Differentiation

III. **Instruction**: The following questions are about your product and process characteristics. Please place a mark at the point on the scale line that matches your initial perception.

13. How does the amount of variety in your products today compare with that of 10 years ago?

Much Less Variety — Much More Variety

14. Are your average product life cycles longer or shorter than they were 10 years ago?

Much Longer ... Much Shorter

15. If possible, please indicate the average life cycle length (in years and/or months) of your products now and 10 years ago:

Product life cycle today = ________ Product life cycle 10 years ago = ____________

16. How does the quality of your products today compare with that of 10 years ago?

Much Lower ... Much Higher

17. How do the production costs of your products today compare with those of 10 years ago?

Much Higher ... Much Lower

18. How do today's products compare with those of 10 years ago in meeting customers' needs?

Much Worse ... Much Better

19. To what extent are your products customized to individual customers?

Not at All ... 100% Customized

20. How does the level of customization in your products today compare with that 10 years ago?

Much Less Customization ... Much More Customization

21. Where does your Production Process lie on the scale between One-of-a-kind production (where each final product is different from others) and fully standardized Mass production?

One-of-a-Kind ... Full Mass Production

22. How much more production flexibility (the ability to change quickly between products) exists in your production process today than 10 years ago? Today's process has ...

Much Flexibility Less | Much Flexibility More

23. Is your product development project longer or shorter than 10 years ago?

Much Longer | Much Shorter

24. If possible, please indicate the average product development project length (in years and/or months) of your products now and 10 years ago:

Today = ___________ 10 years ago = ___________

25. If your company/division is providing more product variety and customization today than 10 years ago, how is it being done? (Check all that apply)

___ Information technology
___ Automation
___ Lean manufacturing
___ (Internet-based) Collaborative engineering
___ Non-value adding reduction (NVAR)
___ Supply chain synchronization
___ Modular manufacturing
___ Others (Please specify) _______________

____

*** Thank you very much for your participation ***

*** The survey results and analyses will be available for participants approximately in May 2003 **

APPENDIX C

Record ID	Survey Date	Industry	Product Types	Company Employees	Division Employees	Public/Private	Company Revenue ($million)	Division Revenue ($million)	NPD-Related Employees	NPD-Related Engineers	P/D Innovation	Competition	Many M&A
001	5/3/2003	2000 - Food & Kindred Products	Consumer products	25000	283	Public	$12,000	$45			Fast changing	High competition	TRUE
002	4/27/2003	3800 - Instruments & Related Products	Industrial products	1900	200	Public	$330	$100	30	20	Slow changing	High competition	FALSE
003	5/12/2003	3500 - Industrial Machinery & Equipment	Industrial products	230	17	Private	$50	$20	7	7	Slow changing	High competition	FALSE
004	4/2/2003	3572 - Computer Storage Devices	One-of-a-kind	45800		Public	$6,087		9000	5000	Fast changing	High competition	FALSE
005	4/9/2003	3625 - Relays and Industrial Controls	Industrial products	140		Private	$120		13	11	Slow changing	High competition	FALSE
006	3/11/2003	3572 - Computer Storage Devices	Industrial products	8898	35	Public	$1,200	$60	8	6	Fast changing	High competition	FALSE
007	2/21/2003	3572 - Computer Storage Devices	Industrial products	45000	3000	Public	$6,000		200	150	Fast changing	High competition	FALSE
008	3/31/2003	3840 - Surgical, Medical, and Dental Instruments	Consumer products	12000	6000	Public	$3,900	$1,000	2000	1000	Fast changing	High competition	TRUE
009	2/26/2003	3489 - Ordnance and Accessories	Consumer products	5300	2000	Public	$1,320	$500	1500	1200	Fast changing	Many M&A	TRUE
010	4/21/2003	3800 - Instruments & Related Products	Industrial products	1600	45	Public	$665	$17	4	4	Slow changing		FALSE
011	3/16/2003	3572 - Computer Storage Devices	Industrial products	9000	60	Public	$1,200	$60	20	12	Fast changing	High competition	FALSE
012	2/25/2003	3489 - Ordnance and Accessories	Consumer products	11000	1000	Public	$2,000	$350	60	30	Slow changing	Many M&A	TRUE
013	4/20/2003	2000 - Food & Kindred Products	One-of-a-kind	20000	700	Public	$27,629	$12,396	10	2		High competition	FALSE
014	3/18/2003	3700 - Transportation Equipment	One-of-a-kind	2000	104	Public	$500	$25	6	1	Fast changing	High competition	FALSE
015	3/20/2003	3489 - Ordnance and Accessories	Consumer products	5300	1600	Public	$1,300		800	400	Fast changing	High competition	FALSE
016	3/29/2003	3600 - Electronic & Other Electrical Equipment	Industrial products	300	8	Private	$70		16	11		High competition	FALSE
017	3/23/2003	3489 - Ordnance and Accessories	Consumer products	5000	2000	Public	$1,320	$500	800	700	Fast changing	High competition	FALSE
018	4/21/2003	3600 - Electronic & Other Electrical Equipment	Industrial products	1000		Public	$350		35	10	Slow changing	High competition	FALSE
019	4/22/2003	3800 - Instruments & Related Products	One-of-a-kind	22900	1000	Public	$4,500		20	15	Slow changing	Many M&A	TRUE
020	4/14/2003	3840 - Surgical, Medical, and Dental Instruments	One-of-a-kind	1000	300	Public	$50		200	75		High competition	TRUE
021	3/14/2003	3625 - Relays and Industrial Controls	Industrial products	600	20	Private	$105		100	70	Fast changing	High competition	FALSE
022	2/25/2003	1500 - General Building Contractors	Industrial products	3000	300	Private	$1,000	$7	14	4	Slow changing	High competition	TRUE
023	5/7/2003	2700 - Printing & Publishing	Consumer products	350	45	Private	$50		5	2	Fast changing	High competition	TRUE
024	3/26/2003	3840 - Surgical, Medical, and Dental Instruments	One-of-a-kind	540	25	Public	$130	$18	32	8	Slow changing		FALSE
025	2/25/2003	3500 - Industrial Machinery & Equipment	Industrial products	71669	150	Public	$17,000		35	20		High competition	FALSE
026	5/20/2003	3500 - Industrial Machinery & Equipment	Industrial products	8000	100	Public	$1,000		250	20	Slow changing	High competition	TRUE
027	4/28/2003	3800 - Instruments & Related Products	Industrial products	180		Private	$20		100	50		No competition	FALSE
028	4/29/2003	3500 - Industrial Machinery & Equipment	Industrial products	95	8	Private	$17		8	6	Fast changing	High competition	FALSE
029	4/22/2003	2830 - Biotechnology & Drugs	Industrial products	80		Private	$15		11	5	Fast changing	High competition	FALSE
030	4/21/2003	2700 - Printing & Publishing	Consumer products	5700		Private	$800		50	15	Slow changing	No competition	FALSE
031	2/26/2003	3840 - Surgical, Medical, and Dental Instruments	One-of-a-kind	71669	500	Public	$17,000	$3,500	100	50	Slow changing	High competition	FALSE
032	2/26/2003	3600 - Electronic & Other Electrical Equipment	Industrial products	2000	25	Public	$350		25	10	Slow changing		FALSE
033	5/8/2003	3600 - Electronic & Other Electrical Equipment	Industrial products	700	7	Private	$65		7	7	Slow changing		FALSE
034	3/7/2003	3600 - Electronic & Other Electrical Equipment	Industrial products	3500	220	Public	$500		800	100	Fast changing	High competition	FALSE
035	3/27/2003	3625 - Relays and Industrial Controls	Industrial products	49000		Public	$8,000	$1,500				High competition	TRUE
036	4/11/2003	3840 - Surgical, Medical, and Dental Instruments	Consumer products	71669		Public	$17,000					High competition	TRUE
037	3/20/2003	3840 - Surgical, Medical, and Dental Instruments	Consumer products	14000	3500	Public	$2,800	$1,000	1000	250	Fast changing	High competition	TRUE
038	2/25/2003	3840 - Surgical, Medical, and Dental Instruments	Consumer products	12000	6000	Public	$3,000	$1,750	4000	2500	Fast changing	High competition	TRUE
039	3/6/2003	3840 - Surgical, Medical, and Dental Instruments	One-of-a-kind	8000	2600	Public	$2,800	$1,000	400	260	Fast changing	High competition	TRUE
040	3/5/2003	2000 - Food & Kindred Products	One-of-a-kind	4500	250	Private	$6,000	$250	250	125	Slow changing	High competition	TRUE
041	4/22/2003	3500 - Industrial Machinery & Equipment	Industrial products	200	200	Private	$35		22	10	Fast changing		FALSE
042	2/17/2003	2700 - Printing & Publishing	Consumer products	8300	100	Public	$1,500	$12	20	0	Fast changing	High competition	FALSE
043	4/24/2003	3500 - Industrial Machinery & Equipment	Industrial products	400		Private	$215		20	4		High competition	FALSE
044	3/28/2003	3840 - Surgical, Medical, and Dental Instruments	Consumer products	20000	650	Public	$6,000	$0	100	50	Fast changing		FALSE
045	3/23/2003	3800 - Instruments & Related Products	Industrial products	63000	30	Public	$22,274		18	9	Fast changing	High competition	FALSE
Average				**14278**	**919**		**3961**	**1096**	**526**	**291**			
Minimum				**80**	**7**		**15**	**0**	**4**	**0**			
Maximum				**71669**	**6000**		**27629**	**12396**	**9000**	**5000**			

Record ID	# Product	Regulatory Agency	P/D Life (yr)	Tech Change	Software	Electronic	Mechanical	Chemical	Foods	Idea Screen	Business Analysis	Development	Testing	Commercialization
001		FDA - Food & Drug Administration	0.5	Incremental change	FALSE	FALSE	FALSE	FALSE	TRUE					
002	100 products	CE - Consumer Electronics Association	10	Incremental change	TRUE	TRUE	TRUE	FALSE	FALSE	80				50
003	10 products	UL - Underwriter Laboratory	5	Platform change	FALSE	TRUE	TRUE	FALSE	FALSE	50	25	15	15	10
004	10 products	FCC - Federal Communications Commission	3	Both	TRUE	TRUE	TRUE	TRUE	FALSE	90	50	30	10	5
005	Legacy products		10	Incremental change	FALSE	TRUE	TRUE	FALSE	FALSE	100	100	95	95	95
006	100 products	UL - Underwriter Laboratory	0.75	Platform change	FALSE	FALSE	TRUE	TRUE	FALSE	95	75	60	40	35
007	Legacy products	FCC - Federal Communications Commission	5	Incremental change	FALSE	TRUE	FALSE	FALSE	FALSE	90	80	80	60	5
008	100 products	FDA - Food & Drug Administration	1	Incremental change	FALSE	FALSE	TRUE	FALSE	FALSE	75	60	50	45	40
009	Legacy products	DOD - Department of Defense	35	Platform change	TRUE	TRUE	TRUE	FALSE	FALSE	80	70	40	20	10
010	Legacy products	NRC - Nuclear Regulatory Commission	25	Incremental change	FALSE	TRUE	FALSE	FALSE	FALSE	50	40	10	10	10
011	10 products		1	Incremental change	FALSE	FALSE	TRUE	FALSE	FALSE	90	90	70	60	40
012	Legacy products	DOD - Department of Defense	6	Incremental change	TRUE	TRUE	TRUE	TRUE	FALSE	25	15	10	5	2
013	100 products	FDA - Food & Drug Administration	10	Incremental change	FALSE	FALSE	FALSE	FALSE	TRUE	100	50	25	25	5
014	10 products	EPA - Environment Protection Agency	10	Platform change	FALSE	FALSE	TRUE	FALSE	FALSE	100	80	60	50	50
015	Legacy products	DOD - Department of Defense	20	Platform change	TRUE	TRUE	TRUE	FALSE	FALSE	75	75	70	70	0
016		FDA - Food & Drug Administration	10	Incremental change	TRUE	TRUE	TRUE	TRUE	FALSE	50	20	8	5	3
017	Legacy products	DOD - Department of Defense	30	Platform change	TRUE	TRUE	TRUE	TRUE	FALSE				60	40
018		UL - Underwriter Laboratory		Incremental change	FALSE	FALSE	TRUE	FALSE	FALSE	75	50	20	20	15
019		FAA - Federal Aviation Administration	17.5	Incremental change	FALSE	TRUE	TRUE	FALSE	FALSE	20	15	10	5	1
020	100 products	FDA - Food & Drug Administration	10	Incremental change	FALSE	FALSE	FALSE	TRUE	FALSE	30	25	25	10	10
021	100 products	FDA - Food & Drug Administration	1	Incremental change	FALSE	TRUE	TRUE	FALSE	FALSE					
022	10 products	NFRC - National Fenestration Rating Council	20	Platform change	FALSE	FALSE	TRUE	FALSE	FALSE	100	80	20	15	2
023	10 products	EPA - Environment Protection Agency	7.5	Incremental change	FALSE	FALSE	TRUE	TRUE	FALSE	50	25	10	8	5
024		FDA - Food & Drug Administration	10	Incremental change	TRUE	TRUE	TRUE	TRUE	FALSE	100			5	2
025	Legacy products	UL - Underwriter Laboratory	5	Incremental change	TRUE	TRUE	TRUE	TRUE	FALSE	10	5	5	5	5
026		EPA - Environment Protection Agency	12.5	Incremental change	FALSE	FALSE	TRUE	TRUE	FALSE					
027		FDA - Food & Drug Administration	6.5	Incremental change	FALSE	TRUE	TRUE	FALSE	FALSE	80	60	30	20	10
028	100 products	UL - Underwriter Laboratory	15	Incremental change	FALSE	TRUE	TRUE	FALSE	FALSE	60	30	15	8	2
029	10 products	FDA - Food & Drug Administration	5	Incremental change	TRUE	TRUE	TRUE	TRUE	FALSE	30	25	15	13	10
030	Legacy products		35	Incremental change	FALSE	FALSE	TRUE	FALSE	FALSE	20	10	8	5	3
031	Legacy products	FDA - Food & Drug Administration	6.5	Incremental change	TRUE	TRUE	TRUE	FALSE	FALSE	95	90	65	50	45
032	Legacy products	UL - Underwriter Laboratory	10	Incremental change	FALSE	TRUE	TRUE	TRUE	FALSE	90	90	90	90	90
033	100 products	FDA - Food & Drug Administration	40	Incremental change	FALSE	TRUE	FALSE	FALSE	FALSE	10	5	3	2	1
034	1 product	EPA - Environment Protection Agency	1.5	Incremental change	FALSE	TRUE	TRUE	FALSE	FALSE	100	90	80	40	35
035	Legacy products	FDA - Food & Drug Administration	11	Incremental change	FALSE	FALSE	TRUE	FALSE	FALSE	50	20	1	1	1
036	Legacy products	FDA - Food & Drug Administration	1	Incremental change	FALSE	FALSE	TRUE	FALSE	FALSE	75	50	40	20	10
037	100 products	FDA - Food & Drug Administration	1.5	Platform change	FALSE	TRUE	TRUE	FALSE	FALSE	99	10	1	1	1
038	100 products	FDA - Food & Drug Administration	5	Incremental change	FALSE	FALSE	TRUE	FALSE	FALSE	75	50	45	40	38
039	100 products	FDA - Food & Drug Administration	2.5	Platform change	FALSE	FALSE	TRUE	FALSE	FALSE	100	99	95	85	80
040	100 products	USDA - US Department of Agricultural	10	Platform change	FALSE	FALSE	FALSE	FALSE	TRUE	60	40	10	5	1
041	100 products	USDA - US Department of Agricultural	8	Platform change	FALSE	FALSE	TRUE	FALSE	FALSE	70	45	15	15	15
042	10 products		3	Incremental change	TRUE	FALSE	FALSE	FALSE	FALSE	50	25	10	5	5
043	100 products	FDA - Food & Drug Administration	5	Both	FALSE	FALSE	FALSE	TRUE	FALSE	80	70	70	60	50
044		FDA - Food & Drug Administration	3	Both	FALSE	TRUE	TRUE	FALSE	FALSE	90	50	20	20	19
045	10 products	FCC - Federal Communications Commission	5	Incremental change	FALSE	TRUE	FALSE	FALSE	FALSE	90	90	68	67	31
Average			**10**							**70**	**51**	**36**	**29**	**21**
Minimum			**0.5**							**10**	**5**	**1**	**1**	**0**
Maximum			**40**							**100**	**100**	**95**	**95**	**95**

Record ID	Industry	Fit customer needs	Lowest product cost	Most innovative features	First-to-market First-to-world	Leverage revenue/profit/market share	Cost reduction	Market repositioning	Addition to existing line	New-to-company	New-to-the-world
001	2000 - Food & Kindred Products	5	3	4	4	4	60	10	20	10	0
002	3800 - Instruments & Related Products	5	1	2	4	3	25	10	30	25	10
003	3500 - Industrial Machinery & Equipment	3	2	2	1	4	15	25	50	10	0
004	3572 - Computer Storage Devices	4	4	5	5	5	20	40	10	10	20
005	3625 - Relays and Industrial Controls	3	1	1	1	2					
006	3572 - Computer Storage Devices	5	4	3	2	5	25	3	50	20	2
007	3572 - Computer Storage Devices	5	5	2	5	5	20	10	0	30	40
008	3840 - Surgical, Medical and Dental Instruments and Supplies	4	4	2	2	3	20	60	10	5	5
009	3489 - Ordnance and Accessories	5	2	4	3	1	10	30	0	0	60
010	3800 - Instruments & Related Products	5	2	4	3	4	10	10	60	10	10
011	3572 - Computer Storage Devices	5	3	4	4	2	15	25	20	15	25
012	3489 - Ordnance and Accessories	5	2	3	0	2	10	10	50	5	25
013	2000 - Food & Kindred Products	4	2	2	4	3	10	5	70	10	5
014	3700 - Transportation Equipment	4	4	5	4	5	8	1	90	1	0
015	3489 - Ordnance and Accessories	4	3	3	1	2	5	5	30	50	10
016	3600 - Electronic & Other Electrical Equipment	4	0	4	3	2	5	0	55	30	10
017	3489 - Ordnance and Accessories	5	2	4		4	30	10	60	0	0
018	3600 - Electronic & Other Electrical Equipment	5	3	4	2	5	25	15	40	15	5
019	3800 - Instruments & Related Products	5	1	1	3	5	5	25	50	15	5
020	3840 - Surgical, Medical and Dental Instruments and Supplies	5	5	3	2	2	30	30	20	10	10
021	3625 - Relays and Industrial Controls	4	2	3	3	4	15	25	30	15	15
022	1500 - General Building Contractors	5	3	4	2	4	45	33	15	5	2
023	2700 - Printing & Publishing	4	4	3	3	4	55	15	10	10	10
024	3840 - Surgical, Medical and Dental Instruments and Supplies	4	0	2	0	0	50	50	0	0	0
025	3500 - Industrial Machinery & Equipment	4	3	4	4	5	10	0	35	45	10
026	3500 - Industrial Machinery & Equipment						35	10	40	10	5
027	3800 - Instruments & Related Products	3	1	2	5	4	5	5	10	10	70
028	3500 - Industrial Machinery & Equipment	4	4	3	2	3	75	5	14	5	1
029	2830 - Biotechnology & Drugs	4	1	3	3	4	10	20	20	30	20
030	2700 - Printing & Publishing	5	3	2	1	4	50	0	35	15	0
031	3840 - Surgical, Medical and Dental Instruments and Supplies	4	4	5	3	4	3	25	60	10	2
032	3600 - Electronic & Other Electrical Equipment	1	2	3	5	4	30	15	40	10	5
033	3600 - Electronic & Other Electrical Equipment	3	3	1	1	2	20	5	75	0	0
034	3600 - Electronic & Other Electrical Equipment	5	1	3	4	3	50	15	20	10	5
035	3625 - Relays and Industrial Controls	5	1	3	3	3	3	7	25	25	40
036	3840 - Surgical, Medical and Dental Instruments and Supplies	3	3	2	2	4	40	0	30	20	10
037	3840 - Surgical, Medical and Dental Instruments and Supplies	5	2	4	2	4	0	20	50	20	10
038	3840 - Surgical, Medical and Dental Instruments and Supplies	4	2	3	2	2	10	15	40	25	10
039	3840 - Surgical, Medical and Dental Instruments and Supplies	5	3	5	2	4	5	0	10	25	60
040	2000 - Food & Kindred Products	4	4	1	1	5	30	40	20	5	5
041	3500 - Industrial Machinery & Equipment	4	3	4	2	4	15	10	15	25	35
042	2700 - Printing & Publishing	4	4	2	2	4	25	25	20	25	5
043	3500 - Industrial Machinery & Equipment	5	4	3	4	3	35	35	10	10	10
044	3840 - Surgical, Medical and Dental Instruments and Supplies	5	1	4	5	5	5	5	55	15	20
045	3800 - Instruments & Related Products	4	2	4	3	5	0	0	0	100	0
Average		4.27	2.57	3.07	2.72	3.55	22.02	16.00	31.68	16.84	13.45
Minimum		1	0	1	0	0	0	0	0	0	0
Maximum		5	5	5	5	5	75	60	90	100	70

Record ID	Industry	Meet profitability	Capture market share	Generate revenue growth	Provide unique benefits	Be innovation to market
001	2000 - Food & Kindred Products	2	2	2	1	3
002	3800 - Instruments & Related Products	2	4	4	4	3
003	3500 - Industrial Machinery & Equipment	3	3	3	4	2
004	3572 - Computer Storage Devices	4	4	4	4	3
005	3625 - Relays and Industrial Controls	3	2	1	2	1
006	3572 - Computer Storage Devices	3	4	4	2	2
007	3572 - Computer Storage Devices	4	4	4	3	3
008	3840 - Surgical, Medical, and Dental Instruments	2	3	2	2	2
009	3489 - Ordnance and Accessories	3	2	1	5	4
010	3800 - Instruments & Related Products	4	4	4	4	4
011	3572 - Computer Storage Devices	5	5	2	4	4
012	3489 - Ordnance and Accessories	1	2	2	3	3
013	2000 - Food & Kindred Products	1	2	2	1	3
014	3700 - Transportation Equipment	5	3	3	4	3
015	3489 - Ordnance and Accessories	3	3	3	5	3
016	3600 - Electronic & Other Electrical Equipment	2	4	3	5	5
017	3489 - Ordnance and Accessories	4	5	4	5	5
018	3600 - Electronic & Other Electrical Equipment	3	2	3	4	3
019	3800 - Instruments & Related Products	3	4	4	4	4
020	3840 - Surgical, Medical, and Dental Instruments	3	3	3	4	3
021	3625 - Relays and Industrial Controls	2	1	3	3	3
022	1500 - General Building Contractors	3	2	2	5	4
023	2700 - Printing & Publishing	2	3	3	4	3
024	3840 - Surgical, Medical, and Dental Instruments	1	1	1	2	3
025	3500 - Industrial Machinery & Equipment	2	2	2	4	4
026	3500 - Industrial Machinery & Equipment	5	4	5	4	4
027	3800 - Instruments & Related Products	1	4	2	3	5
028	3500 - Industrial Machinery & Equipment	3		2	4	2
029	2830 - Biotechnology & Drugs	2	2	3	3	3
030	2700 - Printing & Publishing	2	1	1	2	2
031	3840 - Surgical, Medical, and Dental Instruments	2	3	3	3	4
032	3600 - Electronic & Other Electrical Equipment	2	2	2	2	1
033	3600 - Electronic & Other Electrical Equipment	2	1	1	3	1
034	3600 - Electronic & Other Electrical Equipment	3	4	2	2	2
035	3625 - Relays and Industrial Controls	2	2	2	4	4
036	3840 - Surgical, Medical, and Dental Instruments	2	3	4	3	2
037	3840 - Surgical, Medical, and Dental Instruments	4	3	4	3	4
038	3840 - Surgical, Medical, and Dental Instruments	4	3	3	2	2
039	3840 - Surgical, Medical, and Dental Instruments	4	4	4	5	5
040	2000 - Food & Kindred Products	4	1	2	3	1
041	3500 - Industrial Machinery & Equipment	4	3	3	4	4
042	2700 - Printing & Publishing	3	3	3	4	3
043	3500 - Industrial Machinery & Equipment	4	3	3	2	4
044	3840 - Surgical, Medical, and Dental Instruments	4	4	4	4	4
045	3800 - Instruments & Related Products	4	0	1	5	4
Average		2.91	2.82	2.73	3.40	3.13
Minimum		1	0	1	1	1
Maximum		5	5	5	5	5

Record ID	Industry	D1	D2	D3	D4	D5	D6	D7	D8	D9	D10	D11	D12	D13	D14	D15	D16	D17
001	2000 - Food & Kindred Products	3	2	3	2	2	0	0	0	1	1	3	1	0	4	0	3	4
002	3800 - Instruments & Related Products	3	4	3	3	2	1	1	2	2	1	1	1	2	3	0	3	4
003	3500 - Industrial Machinery & Equipment	3	3	4	5	4	3	3	4	3	3	3	3	2	4	0	4	5
004	3572 - Computer Storage Devices	4	4	5	4	3	3	3	3	4	3	4	3	3	4	3	5	3
005	3625 - Relays and Industrial Controls	1	1	2	2	3	1	1	3	2	2	4	0	2	3	0	3	4
006	3572 - Computer Storage Devices	4	4	4	3	1	1	4	3	1	3	3	1	3	0	2	5	3
007	3572 - Computer Storage Devices	3	4	5	5	5	4	5	5	5	4	5	4	3	2	5	5	3
008	3840 - Surgical, Medical, and Dental Instruments	5	5	4	4	4	3	3	4	3	4	5	3	3	3	2	2	4
009	3489 - Ordnance and Accessories	5	5	4	5	5	5	4	4	4	4	4	5	4	2	4	4	3
010	3800 - Instruments & Related Products	5	4	4	3	3	4	5	4	4	5	5	4	3	3	2	2	3
011	3572 - Computer Storage Devices	2	1	2	2	3	3	4	4	4	5	3	3	5	4	0	5	3
012	3489 - Ordnance and Accessories	5	5	4	4	4	4	5	5	4	4	4	4	3	3	5	5	5
013	2000 - Food & Kindred Products	4	2	3	2	4	1	3	4	4	4	4	2	1	3	1	1	2
014	3700 - Transportation Equipment	0	0	1	1	1	0	0	0	0	0	1	0	0	3	0	0	5
015	3489 - Ordnance and Accessories	5	5	5	5	5	5	5	5	4	5	5	5	3	4	5	5	3
016	3600 - Electronic & Other Electrical Equipment	5	0	2	2	2	1	2	4	5	5	3	3	0	4	0	2	3
017	3489 - Ordnance and Accessories	5	5	5	5	5	5	5	5		5	5	5	5	5	5	4	4
018	3600 - Electronic & Other Electrical Equipment	1	1	3	1	3	0	1	3		2	1	0	1	2		4	3
019	3800 - Instruments & Related Products	3	4	3	4	4	4	2	4	3	5	3	4	4	2	0	3	0
020	3840 - Surgical, Medical, and Dental Instruments	5	3	3	3	5	3	2	0	0	1	5	5	3	5	0	0	5
021	3625 - Relays and Industrial Controls	3	4	3	4	4	3	2	3	3	3	3	3	4	4	3	4	4
022	1500 - General Building Contractors	4	1	1	1	3	1	0	3	5	3	5	2	1	3	0	1	1
023	2700 - Printing & Publishing	3	3	4	4	4	2	2	2	4	3	3	4	3	4	2	4	3
024	3840 - Surgical, Medical, and Dental Instruments	0	0	0	0	1	1	0	1	2	0	0	0	0	3	0	5	4
025	3500 - Industrial Machinery & Equipment	1	1	0	0	2	0	0	3	4	3	2	0	0	4	0	3	1
026	3500 - Industrial Machinery & Equipment	4	3	4	4	3	4	3	2	4	2	3	3	3	4	3	5	4
027	3800 - Instruments & Related Products	4	3	1	2	2	1	1	3	1	2	4	1	4	4	0	4	2
028	3500 - Industrial Machinery & Equipment	3	3	3	3	3	1	0	1	1	2	4	3	4	3	2	2	4
029	2830 - Biotechnology & Drugs	3	4	3	2	4	3	1	2	3	4	3	3	2	3	1	3	3
030	2700 - Printing & Publishing	2	1	1	1	1	1	1	2	0	1	0	0	0	0	0	0	0
031	3840 - Surgical, Medical, and Dental Instruments	5	5	3	4	2	2	0	4	4	5	5	0	2	4	0	1	5
032	3600 - Electronic & Other Electrical Equipment	0	2	1	2	2	2	3	2	0	4	5	3	2	1	4	3	4
033	3600 - Electronic & Other Electrical Equipment	0	0	1	0	1	0	1	1	1	1	3	2	2	2	0	2	3
034	3600 - Electronic & Other Electrical Equipment	5	5	5	5	5	5	5	5	5	5	5	5	5	3	3	5	3
035	3625 - Relays and Industrial Controls	5	5	3	3	3	3	1	2	1	3	5	1	2	4	1	2	4
036	3840 - Surgical, Medical, and Dental Instruments	5	2	1	2	3	2	3	4	3	4	2	3	2	4	1	2	1
037	3840 - Surgical, Medical, and Dental Instruments	5	0	4	4	4	2	2	4	1	4	5	3	2	3	1	1	4
038	3840 - Surgical, Medical, and Dental Instruments	5	5	4	4	4	2	1	2	3	3	3	2	3	2	0	2	4
039	3840 - Surgical, Medical, and Dental Instruments	5	5	5	4	5	4	3	5	5	5	5	2	5	2	1	5	5
040	2000 - Food & Kindred Products	5	1	0	2	4	2	1	2	0	3	1	0	1	1	0	2	3
041	3500 - Industrial Machinery & Equipment	3	3	2	3	3	2	1	3	3	3	3	2	2	3	2	2	3
042	2700 - Printing & Publishing	4	4	4	4	5	4	3	4	5	5	4	5	2	5	2	5	4
043	3500 - Industrial Machinery & Equipment	2	2	2	2	4	3	0	0	3	3	2	1	2	2	3	3	4
044	3840 - Surgical, Medical, and Dental Instruments	5	3	4	4	4	2	4	5	5	4	5	5	4	2	3	2	3
045	3800 - Instruments & Related Products		4	2	1	4	1	4	5	3	2	0	0	4	1	4	4	5
Average		3.45	2.91	2.89	2.89	3.29	2.31	2.22	3.02	2.84	3.18	3.36	2.42	2.47	2.98	1.59	3.04	3.33
Minimum		0	0	0	0	1	0	0	0	0	0	0	0	0	0	0	0	0
Maximum		5	5	5	5	5	5	5	5	5	5	5	5	5	5	5	5	5

Record ID	Industry	QFD	Six-Sigma	Fish-bone Diagram	Taguchi	Industrial Standard	Company	Customer	Others	How quality std change?	Competition Landscape
001	2000 - Food & Kindred Products	1	0	1	0	0	1	0		Increasing	High, risky returns
002	3800 - Instruments & Related Products	0	0	0	0	0	1	1		Increasing	Low, stable returns
003	3500 - Industrial Machinery & Equipment	0	0	0	0	1	1	1		Increasing	Low, stable returns
004	3572 - Computer Storage Devices	1	1	1	1	1	1	1		Increasing	Low, stable returns
005	3625 - Relays and Industrial Controls	0	0	0	0	1	0	1		Stable	Low, stable returns
006	3572 - Computer Storage Devices	1	0	0	0	0	0	1		Vary upon products	Low, stable returns
007	3572 - Computer Storage Devices	1	1	1	0	0	1	1		Increasing	High, risky returns
008	3840 - Surgical, Medical, and Dental Instruments	0	0	1	0	1	1	0		Increasing	High, risky returns
009	3489 - Ordnance and Accessories	1	0	0	0	0	0	1		Increasing	High, risky returns
010	3800 - Instruments & Related Products	1	1	0	0	1	0	0		Increasing	High, risky returns
011	3572 - Computer Storage Devices	0	0	0	0	0	0	1		Increasing	High, stable returns
012	3489 - Ordnance and Accessories	1	1	1	1	0	0	1		Increasing	Low, stable returns
013	2000 - Food & Kindred Products	0	0	0	0	0	1	1		Increasing	Low, stable returns
014	3700 - Transportation Equipment	0	0	0	0	0	1	1		Increasing	Low, stable returns
015	3489 - Ordnance and Accessories	0	0	0	0	0	0	1		Stable	High, stable returns
016	3600 - Electronic & Other Electrical Equipment	1	0	0	0	0	0	0		Decreasing	High, risky returns
017	3489 - Ordnance and Accessories	0	0	0	0	0	0	1		Decreasing	High, stable returns
018	3600 - Electronic & Other Electrical Equipment	0	0	0	0	1	1	1		Stable	Low, stable returns
019	3800 - Instruments & Related Products	0	0	0	0	0	1	1		Increasing	High, stable returns
020	3840 - Surgical, Medical, and Dental Instruments	1	0	0	0	0	1	0		Increasing	High, risky returns
021	3625 - Relays and Industrial Controls	1	0	0	0	1	0	0		Increasing	High, stable returns
022	1500 - General Building Contractors	0	0	0	0	0	1	0		Increasing	Low, stable returns
023	2700 - Printing & Publishing	0	0	0	0	0	1	0		Increasing	High, risky returns
024	3840 - Surgical, Medical, and Dental Instruments	0	0	0	0	1	0	1		Stable	High, risky returns
025	3500 - Industrial Machinery & Equipment	0	1	0	0	0	1	0		Increasing	High, risky returns
026	3500 - Industrial Machinery & Equipment	0	0	1	0	0	1	1		Increasing	Low, risky returns
027	3800 - Instruments & Related Products	1	0	0	0	1	1	1		Vary upon products	High, stable returns
028	3500 - Industrial Machinery & Equipment	0	0	0	0	0	0	0			Low, stable returns
029	2830 - Biotechnology & Drugs	1	0	0	0	0	1	0		Increasing	Low, risky returns
030	2700 - Printing & Publishing	1	0	0	0	0	0	1		Increasing	High, stable returns
031	3840 - Surgical, Medical, and Dental Instruments	1	1	0	0	0	1	1		Increasing	Low, stable returns
032	3600 - Electronic & Other Electrical Equipment	0	0	0	0	0	0	1		Stable	Low, risky returns
033	3600 - Electronic & Other Electrical Equipment	0	0	0	0	1	1	1		Vary upon products	Low, stable returns
034	3600 - Electronic & Other Electrical Equipment	0	0	0	0	0	0	1		Increasing	High, risky returns
035	3625 - Relays and Industrial Controls	0	0	0	0	0	1	0		Increasing	Low, risky returns
036	3840 - Surgical, Medical, and Dental Instruments	1	1	1	0	0	1	1		Increasing	
037	3840 - Surgical, Medical, and Dental Instruments	0	0	0	0	0	1	0	FDA	Stable	High, risky returns
038	3840 - Surgical, Medical, and Dental Instruments	1	1	0	0	0	1	0	FDA	Stable	High, stable returns
039	3840 - Surgical, Medical, and Dental Instruments	1	1	1	0	0	1	0		Increasing	High, stable returns
040	2000 - Food & Kindred Products	1	0	0	0	0	1	0		Increasing	Low, stable returns
041	3500 - Industrial Machinery & Equipment	1	0	0	0	0	1	0		Increasing	High, stable returns
042	2700 - Printing & Publishing	0	0	0	0	0	0	1		Vary upon products	Low, risky returns
043	3500 - Industrial Machinery & Equipment	0	1	0	0	0	1	0		Increasing	Low, stable returns
044	3840 - Surgical, Medical, and Dental Instruments	1	1	1	0	0	1	1	FDA	Increasing	High, risky returns
045	3800 - Instruments & Related Products	1	1	0	0	1	0	1		Vary upon products	High, stable returns
Average		47%	27%	20%	4%	24%	62%	60%			
Minimum											
Maximum											

Record ID	Industry	E1	E2	E3	E4	E5	E6	E7	E8	E9	E10	E11	E12	E13	E14
001	2000 - Food & Kindred Products	2	4	3	2	5	3	3	3	2	3	0	2	5	4
002	3800 - Instruments & Related Products	4	1	1	1	1	3	4	4	3	2	4	3	3	3
003	3500 - Industrial Machinery & Equipment	2	1	5	1	2	2	5	2	2	3	3	1	4	2
004	3572 - Computer Storage Devices	5	4	4	4	4	4	5	4	3	3	3	3	4	4
005	3625 - Relays and Industrial Controls	0	1	0	3	3	2	4	4	5	1	4	2	4	4
006	3572 - Computer Storage Devices	4	5	3	1	1	3	5	4	2	1	3	3	4	1
007	3572 - Computer Storage Devices	5	2	2	3	3	3	5	1	4	2	3	4	4	3
008	3840 - Surgical, Medical, and Dental Instruments	4	4	4	3	5	2	5	4	4	5	4	2	5	4
009	3489 - Ordnance and Accessories	5	4	0	5	1	4	4	0	5	3	5	5	4	4
010	3800 - Instruments & Related Products	4	2	3	3	2	4	3	2	3	2	3	4	3	5
011	3572 - Computer Storage Devices	2	1	1	1	1	3	3	1	1	3	3	2	4	4
012	3489 - Ordnance and Accessories	5	4	3	3	3	3	5	2	3	5	5	4	5	4
013	2000 - Food & Kindred Products	3	2	1	2	2	2	3	2	1	1	2	2	3	2
014	3700 - Transportation Equipment	0	0	0	0	2	2	4	4	0	0	0	0	3	3
015	3489 - Ordnance and Accessories	5	5	2	5	2	3	5	3	3	3	3	5	5	5
016	3600 - Electronic & Other Electrical Equipment	3	2	4	3	3	3	5	5	4	0	0	4	4	3
017	3489 - Ordnance and Accessories	5	5	5	5	5	5	5		5	4	5	5	5	5
018	3600 - Electronic & Other Electrical Equipment	3	2	3	4	2	3	3	2	2	1	1	2	4	4
019	3800 - Instruments & Related Products	5	5		1	4	2	5	5		4	1	5	5	5
020	3840 - Surgical, Medical, and Dental Instruments	5	5	3	5	4	2	5	5	3	0	0	5	5	5
021	3625 - Relays and Industrial Controls	4	4	2	2	3	3	4	3	2	2	3	3	4	3
022	1500 - General Building Contractors	5	4	4	4	2	1	2	1	2	1	1	2	4	4
023	2700 - Printing & Publishing	3	2	1	2	4	4	4	3	2	4	4	4	4	4
024	3840 - Surgical, Medical, and Dental Instruments	4	2	3	2	0	0	1	3	0	0	2	2	5	4
025	3500 - Industrial Machinery & Equipment	4	4	4	3	4	2	4	1	3	1	0	4	5	3
026	3500 - Industrial Machinery & Equipment	5	3	4	3	1	4	5	1	0	3	3	5	4	3
027	3800 - Instruments & Related Products	4	3	5	3	3	2	3	2	2	2	2	2	5	3
028	3500 - Industrial Machinery & Equipment	4	3	1	2	2	4	4			4	1	4	4	4
029	2830 - Biotechnology & Drugs	4	4	3	3	2	2	3	2	2	2	1	3	3	3
030	2700 - Printing & Publishing	3	3	1	1	1	2	4	1	0	0	2	2	4	3
031	3840 - Surgical, Medical, and Dental Instruments	3	2	4	4	4	2	3	3	4	1	1	2	5	4
032	3600 - Electronic & Other Electrical Equipment	4	5	3	3	4	2	4	4	5	2	3	3	3	3
033	3600 - Electronic & Other Electrical Equipment	3	0	1	2	1	1	2	1	1	1	2	2	1	0
034	3600 - Electronic & Other Electrical Equipment	4	4	4	3	4	4	5	5	3	3	5	4	4	4
035	3625 - Relays and Industrial Controls	5	4	4	2	2	2	3	2	2	2	3	3	3	3
036	3840 - Surgical, Medical, and Dental Instruments	2		3	1	4	1	3	1	1	1		2	3	3
037	3840 - Surgical, Medical, and Dental Instruments	3	4	4	2	4	3	3	3	1	4	1	4	4	1
038	3840 - Surgical, Medical, and Dental Instruments	5	5	3	3	4	3	4	4	3	4	3	4	2	3
039	3840 - Surgical, Medical, and Dental Instruments	5	5	5	5	5	4	4	4	3	0	1	4	5	5
040	2000 - Food & Kindred Products	4	3	4	3	5	3	4	4	2	1	2	4	3	2
041	3500 - Industrial Machinery & Equipment	4	3	3	3	3	3	2	2	3	2	2	3	4	3
042	2700 - Printing & Publishing	5	4	4	3	3	4	5	3	4	3	4	4	5	3
043	3500 - Industrial Machinery & Equipment	3	3	2	3	2	3	3	3	3	2	3	3	3	3
044	3840 - Surgical, Medical, and Dental Instruments	4	3	3	4	2	2	4	2	2	2	3	2	4	2
045	3800 - Instruments & Related Products	5	4	1	5	4	2	4	1	3	4	2	3	5	4
Average		3.78	3.18	2.80	2.80	2.84	2.69	3.84	2.70	2.51	2.16	2.41	3.13	3.98	3.36
Minimum		0	0	0	0	0	0	1	0	0	0	0	0	1	0
Maximum		5	5	5	5	5	5	5	5	5	5	5	5	5	5

Record ID	Industry	E15	E16	E17	MS Project	MS Excel	Stage- gate	Squeaky Wheel	Best case	Worst case	Realistic case	E20
001	2000 - Food & Kindred Products	3	5	5	1	0	0	0	0	0	1	Research
002	3800 - Instruments & Related Products	1	1	2	1	0	0	0	0	0	1	Design
003	3500 - Industrial Machinery & Equipment	2	1	1	1	1	0	1	1	0	0	Design
004	3572 - Computer Storage Devices	5	3	4	1	1	1	0	0	0	1	Feasibility
005	3625 - Relays and Industrial Controls	3	0	0	0	0	0	0	0	0	0	Design
006	3572 - Computer Storage Devices	2	5	3	1	1	0	0	0	0	1	Design
007	3572 - Computer Storage Devices	4	3	2	1	1	0	0	0	0	1	Design
008	3840 - Surgical, Medical, and Dental Instruments	3	3	2	1	0	0	0	1	0	0	Design
009	3489 - Ordnance and Accessories	3	3	4	1	0	0	0	1	0	1	Conceptualization
010	3800 - Instruments & Related Products	3	1	2	1	0	0	1	0	0	1	Feasibility
011	3572 - Computer Storage Devices	1	3	2	1	0	0	0	1	0	0	Prototyping
012	3489 - Ordnance and Accessories	3	5	4	1	1	1	1	1	0	0	Feasibility
013	2000 - Food & Kindred Products	2	1	1	1	1	0	0	0	0	1	Prototyping
014	3700 - Transportation Equipment	0	0	1	0	1	0	0	0	0	1	Prototyping
015	3489 - Ordnance and Accessories	4	3	5	1	1	1	0	0	0	1	Conceptualization
016	3600 - Electronic & Other Electrical Equipment	2	2	2	1	0	0	0	0	0	1	Design
017	3489 - Ordnance and Accessories	5	5	5	1	0	0	0	0	1	0	Conceptualization
018	3600 - Electronic & Other Electrical Equipment	2	1	3	1	0	0	1	1	0	0	Design
019	3800 - Instruments & Related Products	3	3	3	1	0	0	0	1	0	0	Feasibility
020	3840 - Surgical, Medical, and Dental Instruments	5	3	4	1	0	0	0	1	0	0	Design
021	3625 - Relays and Industrial Controls	4	4	3	1	0	0	0	0	0	1	Feasibility
022	1500 - General Building Contractors	3	3	1	1	1	1	0	0	0	1	Design
023	2700 - Printing & Publishing	3	2	2	0	1	0	0	0	0	1	Prototyping
024	3840 - Surgical, Medical, and Dental Instruments	0	0	0	1	0	0	0	0	0	0	Design
025	3500 - Industrial Machinery & Equipment	1	3	1	1	1	0	0	1	0	0	Design
026	3500 - Industrial Machinery & Equipment	4	4	4	1	0	0	0	0	0	1	Feasibility
027	3800 - Instruments & Related Products	2	2	2	1	1	0	1	1	0	0	Prototyping
028	3500 - Industrial Machinary & Equipment	4	3	3	0	0	0	0	0	0	0	
029	2830 - Biotechnology & Drugs	2	1	2	1	1	0	0	0	0	1	Research
030	2700 - Printing & Publishing	0	0	0	1	0	1	1	1	0	0	Feasibility
031	3840 - Surgical, Medical, and Dental Instruments	3	1	3	1	1	0	0	1	0	0	Design
032	3600 - Electronic & Other Electrical Equipment	0	0	2	1	1	0	1	0	1	0	Prototyping
033	3600 - Electronic & Other Electrical Equipment	0	0	0	0	1	0	1	1	0	0	Design
034	3600 - Electronic & Other Electrical Equipment	5	5	5	1	0	0	0	0	0	1	Feasibility
035	3625 - Relays and Industrial Controls	2	2	3	1	1	0	0	0	0	1	Design
036	3840 - Surgical, Medical, and Dental Instruments	4	1	0	1	1	1	1	1	0	0	Feasibility
037	3840 - Surgical, Medical, and Dental Instruments	1	2		1	0	0	0	0	0	1	Design
038	3840 - Surgical, Medical, and Dental Instruments	4	3	3	1	1	1	1	0	0	1	Feasibility
039	3840 - Surgical, Medical, and Dental Instruments	5	5	5	1	1	0	0	0	0	1	Feasibility
040	2000 - Food & Kindred Products	3	5	3	1	0	0	0	0	0	1	Feasibility
041	3500 - Industrial Machinery & Equipment	3	4	4	1	0	0	0	0	0	1	Research
042	2700 - Printing & Publishing	4	5	4	1	1	0	0	0	0	1	Conceptualization
043	3500 - Industrial Machinery & Equipment	3	0	1	1	1	0	0	0	0	1	Prototyping
044	3840 - Surgical, Medical, and Dental Instruments	2	2	3	1	0	0	0	1	0	0	Design
045	3800 - Instruments & Related Products	1	3	5	1	0	1	0	1	0	1	Design
Average		2.64	2.47	2.59	89%	49%	18%	22%	36%	4%	58%	
Minimum		0	0	0								
Maximum		5	5	5								

Record ID	Industry	F1	F2	F3	F4	F5	F6	F7	F8	F9	F10	F11	F12	F13	F14
001	2000 - Food & Kindred Products	4	5	4	3	3	3	4	4	2	2	3	4	4	3
002	3800 - Instruments & Related Products	3	1	4	1	3	1	3	2	3	4	4	4	4	4
003	3500 - Industrial Machinery & Equipment	2	1	3	1	2	4	2	2	2	3	3	3	1	1
004	3572 - Computer Storage Devices	5	5	4	4	4	5	5	3	5	4	3	4	3	3
005	3625 - Relays and Industrial Controls	5	5	4	0	5	5	3	2	0	5	5	1	0	4
006	3572 - Computer Storage Devices	1	0	5	5	2	2	1	2	3	2	2	4	5	1
007	3572 - Computer Storage Devices	3	2	3	2	4	5	3	2	4	4	2	3	3	5
008	3840 - Surgical, Medical, and Dental Instruments	3	3	3	2	4	5	4	3	5	5	5	3	5	3
009	3489 - Ordnance and Accessories	3	4	4	3	4	4	5	0	5	5	4	3	4	3
010	3800 - Instruments & Related Products	2	1	3	1	2	2	5	3	3	4	4	3	4	4
011	3572 - Computer Storage Devices	1	1	5	2	2	3	4	2	3	3	2	2	4	2
012	3489 - Ordnance and Accessories	5	3	4	5	4	4	4	5	5	5	4	3	5	4
013	2000 - Food & Kindred Products	2	1	2	0	1	0	1	1	2	1	1	1	0	3
014	3700 - Transportation Equipment	0	0	4	0	2	3	0	3	3	0	2	5	0	2
015	3489 - Ordnance and Accessories	3	3	5	2	5	5	5	5	5	4	4	5	5	4
016	3600 - Electronic & Other Electrical Equipment	4	5	3	1	4	4	5	3	5	5	4	3	0	2
017	3489 - Ordnance and Accessories	5	3	4	5	4	4	5	4	5	5	5	5	5	5
018	3600 - Electronic & Other Electrical Equipment	3	5	3	4	1	0	1	2	4	3	2	3	4	2
019	3800 - Instruments & Related Products	4	1	4	2	4	4	4	3	3	3	1	2		2
020	3840 - Surgical, Medical, and Dental Instruments	3	4	2	0	2	5	5	4	5	5	5	3	4	3
021	3625 - Relays and Industrial Controls	4	5	3	2	2	3	3	3	3	4	3	4	5	4
022	1500 - General Building Contractors	1	0	4	2	4	5	4	4	5	3	3	2	4	4
023	2700 - Printing & Publishing	3	2	2	3	2	2	3	2	3	3	3	3	2	4
024	3840 - Surgical, Medical, and Dental Instruments	5	3	0	0	4	4	0	0	5	5	5	1	2	1
025	3500 - Industrial Machinery & Equipment	3	1	2	2	3	3	2	1	3	2	1	2	1	2
026	3500 - Industrial Machinery & Equipment	3	2	2	3	3	3	3	3	3	3	2	1	5	2
027	3800 - Instruments & Related Products	3	4	2	1	1	2	2	2	2	3	2	1	2	2
028	3500 - Industrial Machinery & Equipment	3	2	2	2	3	0	2	1	3	3	1	2	3	4
029	2830 - Biotechnology & Drugs	3	2	2	2	3	4	4	4	5	4	4	3	2	2
030	2700 - Printing & Publishing	2	2	1	0	2	1	0	1	3	1	0	0	0	1
031	3840 - Surgical, Medical, and Dental Instruments	4	4	3	1	5	3	2	3	5	4	5	2	2	4
032	3600 - Electronic & Other Electrical Equipment	4	0	4	2	1	1	0	0	2	4	4	3	5	3
033	3600 - Electronic & Other Electrical Equipment	0	0	0	0	1	0	0	0	0	0	0	0	0	1
034	3600 - Electronic & Other Electrical Equipment	4	2	4	4	4	4	5	5	5	5	3	3	5	3
035	3625 - Relays and Industrial Controls	4	5	3	2	2	5	2	3	5	5	4	2	1	3
036	3840 - Surgical, Medical, and Dental Instruments	1	4	3	1	1	5	3	4	3	3	2	4	4	3
037	3840 - Surgical, Medical, and Dental Instruments	5	5	3	1		5	3	2	4	5	3	4	5	4
038	3840 - Surgical, Medical, and Dental Instruments	4	4	3	2	3	4	3	3	5	5	5	4	3	3
039	3840 - Surgical, Medical, and Dental Instruments	5	5	2	5	4	5	4	5	5	5	5	5	5	5
040	2000 - Food & Kindred Products	5	4	2	3	3	2	3	2	2	4	3	3	4	5
041	3500 - Industrial Machinery & Equipment	4	4	3	3	3	2	2	3	3	4	3	4	3	3
042	2700 - Printing & Publishing	5	4	4	4	4	0	2	3	3	3	3	3	2	4
043	3500 - Industrial Machinery & Equipment	5	5	4	0	0	2	0	0	1	5	4	4	4	3
044	3840 - Surgical, Medical, and Dental Instruments	4	2	3	2	3	3	4	3	5	5	2	3	4	2
045	3800 - Instruments & Related Products	4	4	2	3	4	4	4		5	4		3	0	2
Average		3.31	2.84	3.02	2.07	2.89	3.11	2.87	2.55	3.56	3.64	3.07	2.89	3.02	2.98
Minimum		0	0	0	0	0	0	0	0	0	0	0	0	0	1
Maximum		5	5	5	5	5	5	5	5	5	5	5	5	5	5

Record ID	Industry	Steering Committee	Peer Review	Phase Review	Expert Review	Periodic Review	CPM / PERT	Cpk	Financial Performance	Time	Quality	Cost	Efficiency	Co-located	Virtually-located	F19
001	2000 - Food & Kindred Products	0	1	0	0	0	1	0	0	0	1	1	1	1	0	Not applicable
002	3800 - Instruments & Related Products	0	1	0	1	1	0	0	0	1	1	1	0	1	0	Not applicable
003	3500 - Industrial Machinery & Equipment	0	1	0	0	0	0	0	1	0	0	0	0	1	0	Not applicable
004	3572 - Computer Storage Devices	0	0	1	0	0	0	1	0	0	0	0	1	1	0	Not applicable
005	3625 - Relays and Industrial Controls	0	0	0	1	0	0	0	0	1	0	1	0	1	0	Not applicable
006	3572 - Computer Storage Devices	0	1	1	0	1	0	1	0	0	1	1	0	1	0	Not applicable
007	3572 - Computer Storage Devices	0	1	1	1	1	0	1	0	1	1	1	0	1	1	Yes
008	3840 - Surgical, Medical, and Dental Instruments	1	0	0	0	0	0	1	1	1	1	1	1	1	0	Not applicable
009	3489 - Ordnance and Accessories	1	1	0	0	0	1	0	0	1	0	1	1	1	1	Yes
010	3800 - Instruments & Related Products	1	1	0	1	1	1	0	0	0	1	0	0	1	0	Not applicable
011	3572 - Computer Storage Devices	0	1	0	0	0	0	1	0	1	1	0	0	0	1	No
012	3489 - Ordnance and Accessories	1	1	1	1	1	0	1	1	1	0	1	0	1	1	Yes
013	2000 - Food & Kindred Products	1	0	0	0	0	0	0	1	0	0	1	0	1	0	Not applicable
014	3700 - Transportation Equipment	0	1	0	0	1	0	0	1	0	1	0	0	1	0	Not applicable
015	3489 - Ordnance and Accessories	1	1	1	1	1	1	0	1	1	0	1	0	1	1	Yes
016	3600 - Electronic & Other Electrical Equipment	0	0	0	0	1	0	0	0	1	1	1	0	1	0	Not applicable
017	3489 - Ordnance and Accessories	0	0	0	0	0	0	0	0	1	1	1	0	1	1	Yes
018	3600 - Electronic & Other Electrical Equipment	1	0	0	0	1	0	0	0	1	0	0	0	1	0	Not applicable
019	3800 - Instruments & Related Products	0	0	0	0	1	1	0	1	1	0	1	0	1	0	Not applicable
020	3840 - Surgical, Medical, and Dental Instruments	0	0	1	0	1	0	0	0	0	1	0	0	1	0	Not applicable
021	3625 - Relays and Industrial Controls	0	0	1	0	0	0	0	1	1	1	1	1	1	0	Not applicable
022	1500 - General Building Contractors	1	0	1	0	1	0	1	1	1	1	1	0	1	0	Not applicable
023	2700 - Printing & Publishing	0	1	0	0	0	0	0	1	0	0	0	1	1	0	Not applicable
024	3840 - Surgical, Medical, and Dental Instruments	0	0	1	0	0	0	0	1	0	0	1	0	1	0	Not applicable
025	3500 - Industrial Machinery & Equipment	1	0	0	0	1	0	0	0	1	0	1	0	1	0	Not applicable
026	3500 - Industrial Machinery & Equipment	0	1	0	0	1	0	0	0	0	0	0	1	1	1	Yes
027	3800 - Instruments & Related Products	1	0	0	0	0	0	1	1	0	1	1	0	1	0	Not applicable
028	3500 - Industrial Machinery & Equipment	0	0	0	0	0	0	0	0	0	0	0	0	1	0	Not applicable
029	2830 - Biotechnology & Drugs	1	1	1	0	1	0	1	1	1	1	1	1	1	0	Not applicable
030	2700 - Printing & Publishing	1	0	1	0	0	0	0	1	1	0	1	1	0	1	No
031	3840 - Surgical, Medical, and Dental Instruments	1	0	0	0	1	0	0	0	1	1	1	0	1	0	Not applicable
032	3600 - Electronic & Other Electrical Equipment	0	0	1	0	1	0	0	0	1	0	1	0	1	0	Not applicable
033	3600 - Electronic & Other Electrical Equipment	0	1	0	0	0	0	0	0	0	0	0	0	1	0	Not applicable
034	3600 - Electronic & Other Electrical Equipment	0	1	0	0	0	0	1	1	1	1	1	1	1	1	Yes
035	3625 - Relays and Industrial Controls	1	0	0	0	1	0	0	0	1	0	1	0	1	0	Not applicable
036	3840 - Surgical, Medical, and Dental Instruments	1	0	0	1	1	1	1	0	1	0	0	0	0	1	No
037	3840 - Surgical, Medical, and Dental Instruments	0	0	1	0	1	0	0	0	1	1	1	0	1	0	Not applicable
038	3840 - Surgical, Medical, and Dental Instruments	0	0	1	1	1	0	1	1	1	1	1	0	1	0	Not applicable
039	3840 - Surgical, Medical, and Dental Instruments	1	1	1	0	1	1	1	1	1	1	1	1	1	0	Not applicable
040	2000 - Food & Kindred Products	0	1	0	0	0	1	0	1	0	1	1	0	1	0	Not applicable
041	3500 - Industrial Machinery & Equipment	0	0	1	0	1	0	0	0	1	0	0	1	1	0	Not applicable
042	2700 - Printing & Publishing	1	1	0	0	1	0	0	1	1	1	1	0	1	1	Yes
043	3500 - Industrial Machinery & Equipment	0	0	1	1	1	0	1	1	1	1	1	1	1	0	Not applicable
044	3840 - Surgical, Medical, and Dental Instruments	1	1	1	1	0	0	1	1	1	1	1	0	1	0	Not applicable
045	3800 - Instruments & Related Products	0	1	1	1	0	1	1	1	1	0	1	0	0	1	Yes
Average		0.40	0.47	0.42	0.24	0.56	0.20	0.36	0.49	0.67	0.53	0.71	0.29	0.91	0.27	
Minimum		0	0	0	0	0	0	0	0	0	0	0	0	0	0	
Maximum		1	1	1	1	1	1	1	1	1	1	1	1	1	1	

Record ID	Industry	Internal Training	External Training	Supplier Training	Industry Conference	F21	F22	Internet	Intranet	Satellite	Teleconference	VPN	MS NetMeetin
001	2000 - Food & Kindred Products	1	1	1	1	12	> 10 years	1	1	0	1	0	
002	3800 - Instruments & Related Products	1	1	0	1	5	1-5 years	1	0	0	1	0	
003	3500 - Industrial Machinery & Equipment	1	1	0	0	5	5-10 years	1	0	0	1	0	
004	3572 - Computer Storage Devices	1	1	0	1	36	5-10 years	1	0	0	1	0	
005	3625 - Relays and Industrial Controls	1	1	0	0	5	1-5 years	0	1	0	1	0	
006	3572 - Computer Storage Devices	1	0	0	0	6	< 1 year	1	0	0	1	0	
007	3572 - Computer Storage Devices	1	1	0	1	50	1-5 years	1	1	0	1	0	
008	3840 - Surgical, Medical, and Dental Instruments	0	1	0	0	50	> 10 years	1	1	0	1	0	
009	3489 - Ordnance and Accessories	1	1	0	1	200	1-5 years	1	1	0	1	1	
010	3800 - Instruments & Related Products	1	1	0	0	4	5-10 years	1	0	0	1	0	
011	3572 - Computer Storage Devices	1	1	0	0	5	1-5 years	1	0	0	0	0	
012	3489 - Ordnance and Accessories	1	1	1	1	8	5-10 years	1	1	0	1	1	
013	2000 - Food & Kindred Products	0	0	0	1	4		1	1	0	1	0	
014	3700 - Transportation Equipment	1	0	1	0	2		1	0	0	0	0	
015	3489 - Ordnance and Accessories	1	1	0	1	20	5-10 years	0	1	1	1	1	
016	3600 - Electronic & Other Electrical Equipment	0	0	0	1	9	1-5 years	0	1	0	1	0	
017	3489 - Ordnance and Accessories	1	1	1	1	250	> 10 years	1	1	0	1	1	
018	3600 - Electronic & Other Electrical Equipment	0	1	0	1	8	1-5 years	1	1	0	1	0	
019	3800 - Instruments & Related Products	1	0	0	0		5-10 years	0	1	0	1	0	
020	3840 - Surgical, Medical, and Dental Instruments	0	1	0	0	5	5-10 years	0	0	0	1	0	
021	3625 - Relays and Industrial Controls	0	1	0	0	8	5-10 years	0	0	0	1	0	
022	1500 - General Building Contractors	1	1	0	0	30	1-5 years	1	0	0	1	0	
023	2700 - Printing & Publishing	1	0	0	0	4	1-5 years	1	1	0	1	0	
024	3840 - Surgical, Medical, and Dental Instruments	1	0	0	0	8	5-10 years	1	1	0	0	0	
025	3500 - Industrial Machinery & Equipment	1	1	1	1	4	5-10 years	1	0	0	1	1	
026	3500 - Industrial Machinery & Equipment	1	1	0	0	7	1-5 years	0	0	0	0	0	
027	3800 - Instruments & Related Products	1	1	0	1	25	< 1 year	1	1	0	1	0	
028	3500 - Industrial Machinery & Equipment	0	0	0	0			0	0	0	0	0	
029	2830 - Biotechnology & Drugs	1	1	1	1	8	1-5 years	1	0	0	1	0	
030	2700 - Printing & Publishing	1	1	1	1	8	1-5 years	0	1	0	1	1	
031	3840 - Surgical, Medical, and Dental Instruments	1	1	0	0	8	5-10 years	1	1	0	1	0	
032	3600 - Electronic & Other Electrical Equipment	0	1	1	1	8	> 10 years	1	1	0	1	0	
033	3600 - Electronic & Other Electrical Equipment	1	1	0	1	3		1	1	0	1	0	
034	3600 - Electronic & Other Electrical Equipment	1	0	0	0	10	5-10 years	1	1	0	1	0	
035	3625 - Relays and Industrial Controls	1	1	0	1	3	1-5 years	1	0	0	1	0	
036	3840 - Surgical, Medical, and Dental Instruments	1	0	0	0	5	5-10 years	1	1	1	1	0	
037	3840 - Surgical, Medical, and Dental Instruments	1	1	0	1	30	1-5 years	0	1	0	0	0	
038	3840 - Surgical, Medical, and Dental Instruments	1	1	0	1	35	1-5 years	0	1	0	1	0	
039	3840 - Surgical, Medical, and Dental Instruments	1	1	0	0	12	1-5 years	1	1	0	1	0	
040	2000 - Food & Kindred Products	1	1	1	1	60	1-5 years	1	1	0	0	0	
041	3500 - Industrial Machinery & Equipment	1	1	0	0	5	1-5 years	0	1	0	0	0	
042	2700 - Printing & Publishing	1	1	1	1	11	< 1 year	1	1	0	0	0	
043	3500 - Industrial Machinery & Equipment	1	1	1	1	5	1-5 years	1	0	0	0	0	
044	3840 - Surgical, Medical, and Dental Instruments	1	1	1	1	6	5-10 years	1	1	0	1	0	
045	3800 - Instruments & Related Products	1	1	1	1	30	1-5 years	1	0	0	1	0	
Average		0.82	0.78	0.29	0.56	23.65		0.73	0.62	0.04	0.78	0.13	0.2
Minimum		0	0	0	0	2		0	0	0	0	0	
Maximum		1	1	1	1	250		1	1	1	1	1	

Record ID	Industry	G1	G2	G3	G4	G5	G6	Internal Expertise	External Expertise	Market vs. Technical Risk
001	2000 - Food & Kindred Products	4	3	4	4	4	4	1	1	Both are at the same level
002	3800 - Instruments & Related Products	2	1	1	4	3	3	1	1	Technical risk is higher
003	3500 - Industrial Machinery & Equipment	0	1	0	4	3	0	1	1	Market risk is higher
004	3572 - Computer Storage Devices	4	4	5	2	4	4	1	0	Both are at the same level
005	3625 - Relays and Industrial Controls	0	3	0	4	3	4	1	1	Technical risk is higher
006	3572 - Computer Storage Devices	0	1	2	4	4	4	1	0	Market risk is higher
007	3572 - Computer Storage Devices	3	5	4	4	4	4	1	0	Technical risk is higher
008	3840 - Surgical, Medical, and Dental Instruments	2	4	3	4	3	4	1	0	Both are at the same level
009	3489 - Ordnance and Accessories	4	4	4	2	4	4	1	1	Technical risk is higher
010	3800 - Instruments & Related Products	4	5	4	4	3	4	1	0	Technical risk is higher
011	3572 - Computer Storage Devices	1	2	2	2	2	3	1	0	Technical risk is higher
012	3489 - Ordnance and Accessories	3	4	4	4	4	3	1	1	Technical risk is higher
013	2000 - Food & Kindred Products	3	1	4	4	2	4	1	0	Market risk is higher
014	3700 - Transportation Equipment	0	0	0	4	4	1	0	1	Both are at the same level
015	3489 - Ordnance and Accessories	3	5	5	3	3	5	1	1	Technical risk is higher
016	3600 - Electronic & Other Electrical Equipment	5	4	0	3	2	4	1	0	Both are at the same level
017	3489 - Ordnance and Accessories		4	5	3	5	3	1	1	Technical risk is higher
018	3600 - Electronic & Other Electrical Equipment	2	1	3	3	3	2	1	1	Market risk is higher
019	3800 - Instruments & Related Products	2	2	4	3	2	2	1	1	Both are at the same level
020	3840 - Surgical, Medical, and Dental Instruments	5	5	5	5	5	5	1	0	Both are at the same level
021	3625 - Relays and Industrial Controls	3	3	4	3	4	3	0	1	Market risk is higher
022	1500 - General Building Contractors	5	4	5	1	4	4	1	1	Both are at the same level
023	2700 - Printing & Publishing	3	3	3	4	4	4	1	0	Both are at the same level
024	3840 - Surgical, Medical, and Dental Instruments	4	5	4	4	4	4	1	0	Technical risk is higher
025	3500 - Industrial Machinery & Equipment	1	1	4	3	1	3	1	0	Both are at the same level
026	3500 - Industrial Machinery & Equipment	2	2	5	5	4	5	1	0	Both are at the same level
027	3800 - Instruments & Related Products	2	1	4	0	2	2	1	1	Market risk is higher
028	3500 - Industrial Machinery & Equipment	2	0	4	3	4	2	0	0	
029	2830 - Biotechnology & Drugs	5	4	4	4	3	3	1	0	Market risk is higher
030	2700 - Printing & Publishing	0	1	1	3	4	2	1	0	Market risk is higher
031	3840 - Surgical, Medical, and Dental Instruments	4	4	2	4	4	2	1	1	Market risk is higher
032	3600 - Electronic & Other Electrical Equipment	4	4	3	2	2	3	1	0	Market risk is higher
033	3600 - Electronic & Other Electrical Equipment	0	2	1	4	2	4	1	1	Technical risk is higher
034	3600 - Electronic & Other Electrical Equipment	5	4	5	3	5	2	1	1	Market risk is higher
035	3625 - Relays and Industrial Controls	1	3	3	3	4	4	1	0	Market risk is higher
036	3840 - Surgical, Medical, and Dental Instruments	3	4	4	2	3	4	1	0	Market risk is higher
037	3840 - Surgical, Medical, and Dental Instruments	4	4	4	2	4	5	1	1	Technical risk is higher
038	3840 - Surgical, Medical, and Dental Instruments	4	4	3	2	4	4	1	1	Technical risk is higher
039	3840 - Surgical, Medical, and Dental Instruments	5	5	5	3	4	4	1	0	Technical risk is higher
040	2000 - Food & Kindred Products	5	4	3	4	2	4	1	0	Market risk is higher
041	3500 - Industrial Machinery & Equipment	3	3	3	3	2	3	1	0	Both are at the same level
042	2700 - Printing & Publishing	2	3	2	4	4	3	1	0	Both are at the same level
043	3500 - Industrial Machinery & Equipment	0	1	4	2	3	3	1	1	Both are at the same level
044	3840 - Surgical, Medical, and Dental Instruments	4	2	5	3	2	3	1	1	Both are at the same level
045	3800 - Instruments & Related Products	0	2	4	2	3	3	1	1	Market risk is higher
Average		2.68	2.93	3.29	3.18	3.31	3.33	93%	49%	
Minimum		0	0	0	0	1	0			
Maximum		5	5	5	5	5	5			

Record ID	Industry	H1	H2	H3	H4	H5	H6	Executiv	Employ	Custom	Supplier	Skill	Senior	Availab	Recognit	Financial rewards	Non-finacial rewards	H10
001	2000 - Food & Kindred Products	5	5	5	5	4	5	1	1	1	0	1	0	0	1	1	0	4
002	3800 - Instruments & Related Products	4	3	2	3	3	5	0	1	1	0	1	1	1	1	0	0	50
003	3500 - Industrial Machinery & Equipment	3	3	2	4	2	5	0	1	1	0	0	0	1	0	0	0	5
004	3572 - Computer Storage Devices	5	4	4	5	4	4	0	1	1	1	1	1	0	1	0	0	5
005	3625 - Relays and Industrial Controls	5	0	0	5	2	5	1	1	1	0	0	0	1	0	1	0	6
006	3572 - Computer Storage Devices	4	2	0	5	1	4	0	1	1	0	1	0	0	0	0	1	1
007	3572 - Computer Storage Devices	5	2	4	4	4	5	0	1	0	0	1	0	1	1	1	1	4
008	3840 - Surgical, Medical, and Dental Instruments	4	3	3	5	5	3	0	1	1	0	1	1	1	0	1	0	20
009	3489 - Ordnance and Accessories	4	4	2	5	4	4	1	1	1	0	1	0	0	1	1	0	20
010	3800 - Instruments & Related Products	4	3	3	4	4	5	1	1	1	0	1	0	1	1	0	1	2
011	3572 - Computer Storage Devices	2	1	1	4	3	4	0	1	1	0	0	0	1	1	0	0	6
012	3489 - Ordnance and Accessories	5	3	3	5	3	4	0	0	1	0	1	0	0	1	0	0	25
013	2000 - Food & Kindred Products	1	0	1	2	3	2	1	0	0	0	0	0	1	1	0	0	0
014	3700 - Transportation Equipment	0	0	0	4	5	1	1	1	1	1	0	0	1	0	0	0	0
015	3489 - Ordnance and Accessories	4	3	4	5	5	5	0	0	1	0	1	0	0	1	1	1	100
016	3600 - Electronic & Other Electrical Equipment	5	0	3	5	3	5	1	1	1	0	1	0	1	1	0	0	6
017	3489 - Ordnance and Accessories	5	2	2	5	3	4	0	1	1	1	1	0	1	1	1	0	50
018	3600 - Electronic & Other Electrical Equipment	4	2	3	4	4	2	1	1	1	0	0	0	1	1	0	0	5
019	3800 - Instruments & Related Products	5			3	4	4	1	1	0	0	0	0	1	0	0	0	10
020	3840 - Surgical, Medical, and Dental Instruments	5	3		2	5	5	0	1	0	0	1	0	0	0	1	0	5
021	3625 - Relays and Industrial Controls	4	3	3	4	3	3	0	0	1	0	0	1	0	0	1	0	15
022	1500 - General Building Contractors	4	1	1	4	3	4	1	1	1	0	0	0	0	0	0	0	2
023	2700 - Printing & Publishing	2	3	1	4	3	4	1	1	1	0	1	1	0	1	0	0	3
024	3840 - Surgical, Medical, and Dental Instruments	0	3	0	5	2	5	1	0	0	0	1	1	1	0	0	0	0
025	3500 - Industrial Machinery & Equipment	1	1	2	3	3	5	0	1	1	0	1	0	1	1	1	0	5
026	3500 - Industrial Machinery & Equipment	3	3	3	4	5	5	0	1	0	0	0	0	1	1	0	0	7
027	3800 - Instruments & Related Products	3	0	1	4	5	3	1	1	1	0	0	1	0	1	1	0	3
028	3500 - Industrial Machinery & Equipment	5	0	0	1	5	4	0	0	0	0	0	0	0	0	0	0	
029	2830 - Biotechnology & Drugs	5	2	2	4	3	4	1	1	1	0	1	0	1	1	1	1	5
030	2700 - Printing & Publishing	5	0	4	4	3	2	1	1	1	0	1	0	1	1	0	0	5
031	3840 - Surgical, Medical, and Dental Instruments	2	2	1	5	4	4	0	1	1	0	0	0	1	0	0	1	11
032	3600 - Electronic & Other Electrical Equipment	4	2	2	3	3	2	0	0	1	0	1	0	0	1	0	0	5
033	3600 - Electronic & Other Electrical Equipment	0	0	0	0	0	3	1	1	1	0	1	0	1	0	0	0	0
034	3600 - Electronic & Other Electrical Equipment	5	3	3	4	2	4	0	1	1	0	1	0	1	1	0	0	10
035	3625 - Relays and Industrial Controls	2	2	2	4	4	4	1	1	1	0	1	0	0	1	0	0	
036	3840 - Surgical, Medical, and Dental Instruments	1	1	3	4	3	2	1	1	1	1	0	1	0	1	1	0	1
037	3840 - Surgical, Medical, and Dental Instruments	5	3	5	5	5	5	1	1	1	0	1	0	0	1	1	0	70
038	3840 - Surgical, Medical, and Dental	4	3	2	5	3	4	0	1	1	0	1	0	1	1	1	0	40

Record ID	Industry	I1	I2	I3	I4	I5	I6
001	2000 - Food & Kindred Products	5	4	2	4	2	5
002	3800 - Instruments & Related Products	1	1	2	4	1	4
003	3500 - Industrial Machinery & Equipment	1	1	1	1	1	3
004	3572 - Computer Storage Devices	3	3	3	3	4	5
005	3625 - Relays and Industrial Controls	0	0	5	5	5	5
006	3572 - Computer Storage Devices	0	2	2	4	1	5
007	3572 - Computer Storage Devices	2	3	2	2	1	5
008	3840 - Surgical, Medical, and Dental Instruments	3	2	4	3	3	5
009	3489 - Ordnance and Accessories	3	4	5	4	4	5
010	3800 - Instruments & Related Products	1	2	2	4	3	5
011	3572 - Computer Storage Devices	2	2	2	2	2	5
012	3489 - Ordnance and Accessories	2	2	4	2	2	5
013	2000 - Food & Kindred Products	1	0	1	1	4	4
014	3700 - Transportation Equipment	0	0	0	0	0	2
015	3489 - Ordnance and Accessories	5	4	5	4	5	5
016	3600 - Electronic & Other Electrical Equipment	3	1	2	2	2	5
017	3489 - Ordnance and Accessories	4	4	4	4	5	5
018	3600 - Electronic & Other Electrical Equipment	1	0	0	2	0	4
019	3800 - Instruments & Related Products	0	1	1	0	2	1
020	3840 - Surgical, Medical, and Dental Instruments	2	5	5	5	5	5
021	3625 - Relays and Industrial Controls	4	3	3	2	2	5
022	1500 - General Building Contractors	1	2	2	1	1	5
023	2700 - Printing & Publishing	3	2	2	3	2	3
024	3840 - Surgical, Medical, and Dental Instruments	2	0	0	0	0	5
025	3500 - Industrial Machinery & Equipment	0	1	3	2	1	4
026	3500 - Industrial Machinery & Equipment	2	1	3	4	3	5
027	3800 - Instruments & Related Products	1	2	2	3	0	3
028	3500 - Industrial Machinery & Equipment	4	2	1	3	1	3
029	2830 - Biotechnology & Drugs	3	3	4	4	5	4
030	2700 - Printing & Publishing	1	0	0	0	2	5
031	3840 - Surgical, Medical, and Dental Instruments	2	1	0	1	2	2
032	3600 - Electronic & Other Electrical Equipment	1	1	0	0	0	4
033	3600 - Electronic & Other Electrical Equipment	0	1	3	1	0	2
034	3600 - Electronic & Other Electrical Equipment	1	3	2	2	1	4
035	3625 - Relays and Industrial Controls	2	4	2	2	2	5
036	3840 - Surgical, Medical, and Dental Instruments	4	1	5	3	1	5
037	3840 - Surgical, Medical, and Dental Instruments	4	1	4	3	2	5
038	3840 - Surgical, Medical, and Dental Instruments	3	3	4	4	2	4
039	3840 - Surgical, Medical, and Dental Instruments	5	1	3	4	5	5
040	2000 - Food & Kindred Products	3	4	3	3	2	5
041	3500 - Industrial Machinery & Equipment	2	2	2	2	3	4
042	2700 - Printing & Publishing	2	1	2	2	2	3
043	3500 - Industrial Machinery & Equipment	2	1	2	3	1	4
044	3840 - Surgical, Medical, and Dental Instruments	0	0	1	0	0	4
045	3800 - Instruments & Related Products	3	1	1	3	0	4
Average		**2.09**	**1.82**	**2.36**	**2.47**	**2.04**	**4.22**
Minimum		**0**	**0**	**0**	**0**	**0**	**1**
Maximum		**5**	**5**	**5**	**5**	**5**	**5**

Record ID	Industry	L9_1	L9_2	L10_1	L10_2	L11_1	L11_2	L12_1	L12_2	L13	L14	L15_1 (yr)	L15_2 (yr)	L16	L17	L18	L19
001	2000 - Food & Kindred Products		10%		20%		80%		45%	100%	100%			100%	65%	95%	80%
002	3800 - Instruments & Related Products	60%	60%	85%	85%	60%	70%	50%	50%	60%	50%	10	10	50%	35%	60%	85%
003	3500 - Industrial Machinery & Equipment	20%	30%	75%	80%	50%	80%	50%	20%	70%	70%	3	5	40%	60%	50%	90%
004	3572 - Computer Storage Devices	20%	90%	20%	90%	15%	95%	45%	55%	80%	85%	0.5	3	85%	95%	90%	80%
005	3625 - Relays and Industrial Controls	50%	50%	40%	70%	80%	80%	60%	40%	50%	50%	10	10	50%	70%	70%	30%
006	3572 - Computer Storage Devices	45%	5%	45%	5%	15%	5%	5%	5%	5%	5%	0.75	4	5%	0%	0%	5%
007	3572 - Computer Storage Devices	65%	65%	25%	85%	30%	100%	40%	10%	75%	90%	0.75	0.5	85%	5%	85%	0%
008	3840 - Surgical, Medical, and Dental Instruments	75%	75%	75%	80%	95%	95%	85%	85%	50%	65%	2	2.5	50%	50%	50%	35%
009	3489 - Ordnance and Accessories	60%	80%	80%	100%	60%	80%	95%	70%	40%	30%	37.5	25	60%	10%	70%	100%
010	3800 - Instruments & Related Products	50%	70%	30%	20%	20%	30%	60%	80%	60%	30%	20	15	65%	65%	80%	100%
011	3572 - Computer Storage Devices	70%	75%	55%	60%	60%	65%	50%	50%	60%	70%	125	2	75%	80%	60%	85%
012	3489 - Ordnance and Accessories	50%	0%	100%	0%	80%	0%	10%	0%	80%	60%	8	12	90%	10%	90%	80%
013	2000 - Food & Kindred Products	30%	40%	20%	30%	45%	80%	40%	30%	95%	30%	1	0.5	90%	50%	75%	70%
014	3700 - Transportation Equipment	30%	55%	60%	40%	60%	90%	30%	15%	60%	70%	3	5	80%	70%	65%	10%
015	3489 - Ordnance and Accessories	20%	70%	10%	90%	25%	80%	10%	50%	85%	80%	25	30	80%	70%	70%	100%
016	3600 - Electronic & Other Electrical Equipment	50%	80%	50%	80%	10%	90%	80%	60%	100%	70%			80%	70%	100%	10%
017	3489 - Ordnance and Accessories	80%	65%	90%	90%	20%	65%	50%	50%	15%	50%	30	30	60%	50%	50%	80%
018	3600 - Electronic & Other Electrical Equipment	50%	50%	30%	60%	30%	70%	50%	50%	90%	60%			70%	70%	80%	40%
019	3800 - Instruments & Related Products	40%	50%	60%	80%	20%	30%	60%	80%	80%	50%	15	20	65%	70%	50%	100%
020	3840 - Surgical, Medical, and Dental Instruments	0%	0%	80%	95%	80%	100%	50%	55%	45%	40%	10	10	100%	25%	70%	80%
021	3625 - Relays and Industrial Controls	50%	65%	50%	35%	50%	55%	50%	50%	75%	65%	1	2	75%	75%	85%	65%
022	1500 - General Building Contractors	50%	50%	40%	80%	20%	85%	90%	35%	95%	80%	3.5	35	60%	65%	85%	100%
023	2700 - Printing & Publishing	0%	0%	5%	5%	30%	40%	45%	0%	10%	0%	0375	1.5	5%	5%	5%	5%
024	3840 - Surgical, Medical, and Dental Instruments	70%	30%	90%	90%	50%	90%	90%	50%	50%	50%	10	10	75%	70%	60%	10%
025	3500 - Industrial Machinery & Equipment	70%	70%	20%	60%	60%	75%	60%	50%	85%	75%	5	8	70%	75%	75%	65%
026	3500 - Industrial Machinery & Equipment	25%	90%	65%	95%	30%	100%	40%	65%	80%	70%	3.5	12.5	100%	50%	85%	70%
027	3800 - Instruments & Related Products	60%	60%	65%	65%	30%	15%	30%	30%	60%	60%	5	8	65%	40%	50%	60%
028	3500 - Industrial Machinery & Equipment	60%	70%	45%	60%	40%	65%	60%	35%	85%	20%	15	10	70%	60%	70%	75%
029	2830 - Biotechnology & Drugs	25%	5%	5%	5%	25%	5%	25%	15%	20%	0%	5	7	5%	5%	0%	15%
030	2700 - Printing & Publishing	15%	35%	30%	50%	55%	70%	50%	55%	85%	70%	15	25	25%	45%	20%	85%
031	3840 - Surgical, Medical, and Dental Instruments	40%	75%	30%	60%	55%	85%	25%	40%	70%	65%	5	17.5	50%	30%	60%	50%
032	3600 - Electronic & Other Electrical Equipment	20%		80%		70%		30%		70%	70%	10	20	70%	70%	75%	70%
033	3600 - Electronic & Other Electrical Equipment	60%	55%	55%	65%	50%	60%	45%	45%	70%	50%	5	5	70%	60%	50%	75%
034	3600 - Electronic & Other Electrical Equipment	40%	70%	90%	80%	100%	90%	30%	60%	70%	80%	1.5	3.5	90%	80%	95%	80%
035	3625 - Relays and Industrial Controls	80%	80%	60%	80%	70%	90%	40%	30%	80%	70%	10	20	60%	80%	70%	80%
036	3840 - Surgical, Medical, and Dental Instruments		10%		50%		90%		20%	80%	70%	3	5		90%	70%	90%
037	3840 - Surgical, Medical, and Dental Instruments	80%	70%	25%	50%	85%	85%	80%	60%	80%	65%	1.5	2	70%	80%	80%	20%
038	3840 - Surgical, Medical, and Dental Instruments	55%	60%	55%	60%	75%	75%	70%	55%	65%	55%	2.5	2.5	50%	60%	60%	25%
039	3840 - Surgical, Medical, and Dental Instruments	100%	100%	55%	65%	100%	100%	65%	70%	100%	80%	1.5	2.5	90%	40%	95%	80%
040	2000 - Food & Kindred Products	40%	60%	20%	75%	40%	100%	30%	65%	95%	85%	10	25	55%	85%	35%	5%
041	3500 - Industrial Machinery & Equipment	90%	85%	70%	55%	50%	50%	70%	75%	85%	50%	8	8	95%	50%	80%	80%
042	2700 - Printing & Publishing	70%		50%		90%		50%		90%	80%	3	7	70%	70%	90%	70%
043	3500 - Industrial Machinery & Equipment	55%	55%	35%	35%	55%	75%	15%	15%	75%	45%	5	3	85%	70%	75%	30%
044	3840 - Surgical, Medical, and Dental Instruments	60%	80%	85%	60%	70%	100%	95%	100%	80%	95%	3	6.5	60%	30%	40%	50%
045	3800 - Instruments & Related Products		80%		80%		15%		40%			6					10%
Average		50%	55%	51%	61%	51%	70%	50%	46%	69%	59%	7527	10.51	66%	55%	65%	58%
Minimum		0%	0%	5%	0%	10%	0%	5%	0%	5%	0%	0375	0.5	5%	0%	0%	0%
Maximum		100%	100%	100%	100%	100%	100%	95%	100%	100%	100%	37.5	35	100%	95%	100%	100%

Record ID	Industry	L20	L21	L22	L23	L24_1 (mnt)	L24_2 (mnt)	IT	Automation	Lean Mnfg	Collaborative Eng	NVAR	SCM	Modular Mnfg
001	2000 - Food & Kindred Products	70%	60%	60%	80%			0	1	0	0	0	1	0
002	3800 - Instruments & Related Products	50%	15%	55%	60%	12	12	1	0	0	0	0	1	0
003	3500 - Industrial Machinery & Equipment	50%	80%	80%	20%	24	15	1	1	1	0	0	0	1
004	3572 - Computer Storage Devices	80%	55%	75%	80%	30	72	1	1	1	0	0	1	0
005	3625 - Relays and Industrial Controls	50%	80%	90%	80%	6	1	0	0	0	0	0	0	1
006	3572 - Computer Storage Devices	5%	5%	5%	0%	0.5	2	1	1	0	0	0	0	0
007	3572 - Computer Storage Devices	50%	100%	60%	30%	14	24	0	0	0	0	0	0	0
008	3840 - Surgical, Medical, and Dental Instruments	25%	85%	40%	20%	18	9	0	0	1	0	0	0	1
009	3489 - Ordnance and Accessories	35%	55%		70%	108	84	0	0	0	0	0	0	0
010	3800 - Instruments & Related Products	75%	20%	80%	50%	48	60	0	0	1	0	1	1	1
011	3572 - Computer Storage Devices	55%	70%	60%	50%	8	12	0	0	0	0	0	0	1
012	3489 - Ordnance and Accessories	70%	20%	60%	60%	48	72	1	0	1	1	0	0	0
013	2000 - Food & Kindred Products	55%	90%	75%	60%	6	12	0	0	0	0	0	0	1
014	3700 - Transportation Equipment	45%	100%	60%	70%	2	5	0	0	1	0	0	0	0
015	3489 - Ordnance and Accessories	55%	50%	55%	70%	60	120	1	1	0	0	1	0	0
016	3600 - Electronic & Other Electrical Equipment	60%	70%	90%	90%			1	1	0	0	0	1	0
017	3489 - Ordnance and Accessories	50%	50%	50%	70%	84	120	1	0	0	0	0	0	1
018	3600 - Electronic & Other Electrical Equipment	90%	20%	50%	0%			0	0	1	0	0	0	1
019	3800 - Instruments & Related Products	50%	30%	50%	60%	12	18	0	0	0	0	0	1	0
020	3840 - Surgical, Medical, and Dental Instruments	90%	90%	45%	10%			0	1	1	0	0	0	0
021	3625 - Relays and Industrial Controls	85%	65%	75%	75%	12	24	0	1	0	0	0	0	1
022	1500 - General Building Contractors	100%	10%	95%	55%	24	60	1	1	1	0	1	1	0
023	2700 - Printing & Publishing	5%	5%	5%	10%	6	12	1	0	0	0	0	0	0
024	3840 - Surgical, Medical, and Dental Instruments	50%	90%	50%	50%	24	24	0	0	0	0	0	0	0
025	3500 - Industrial Machinery & Equipment	70%		65%	65%	6	12	1	1	0	0	0	1	0
026	3500 - Industrial Machinery & Equipment	85%	75%	80%	85%	8	48	0	1	0	0	1	0	0
027	3800 - Instruments & Related Products	40%	10%	70%	70%	12	36	0	0	0	0	0	0	1
028	3500 - Industrial Machinery & Equipment	70%	45%	65%	65%	7	12	1	0	0	1	0	1	1
029	2830 - Biotechnology & Drugs	5%	25%	5%		12	12	0	1	0	0	0	1	0
030	2700 - Printing & Publishing	90%	10%	95%	50%			1	1	1	0	0	0	0
031	3840 - Surgical, Medical, and Dental Instruments	60%	95%	65%	20%	36	18	0	1	1	0	1	0	0
032	3600 - Electronic & Other Electrical Equipment	75%	70%	70%	70%	8	18	0	1	1	0	0	0	0
033	3600 - Electronic & Other Electrical Equipment	60%	25%	60%	70%	36	24	1	1	0	0	0	0	0
034	3600 - Electronic & Other Electrical Equipment	40%	80%	50%	90%	6	12	0	0	0	0	0	0	0
035	3625 - Relays and Industrial Controls	60%	60%	70%	70%	18	24	1	1	1	0	1	1	0
036	3840 - Surgical, Medical, and Dental Instruments	100%	70%	60%	60%	24	60	1	0	0	0	0	1	0
037	3840 - Surgical, Medical, and Dental Instruments	15%	70%	75%	60%	18	24	0	0	1	0	1	1	0
038	3840 - Surgical, Medical, and Dental Instruments	20%	65%	60%	30%	24	12	0	0	0	0	0	0	0
039	3840 - Surgical, Medical, and Dental Instruments	65%	90%	85%	90%	30	15	1	1	1	0	1	1	1
040	2000 - Food & Kindred Products	50%	100%	85%	80%	18	23	1	1	1	0	1	1	0
041	3500 - Industrial Machinery & Equipment	60%	50%	25%	75%	6	10	1	1	0	0	1	1	1
042	2700 - Printing & Publishing	90%	75%	80%	80%	6	12	1	1	1	1	0	0	0
043	3500 - Industrial Machinery & Equipment	45%	80%	30%	45%	12	12	1	1	1	0	0	1	0
044	3840 - Surgical, Medical, and Dental Instruments	85%	60%	75%	65%	18	24	1	0	0	0	0	0	0
045	3800 - Instruments & Related Products		50%			18	24	0	0	0	0	0	0	0
Average		58%	57%	61%	57%	21.74	29.75	49%	49%	40%	7%	22%	38%	29%
Minimum		5%	5%	5%	0%	0.5	1							
Maximum		100%	100%	95%	90%	108	120							

References

Adams, M.E., Day, G.S., Dougherty, D. (1998). Enhancing NPD performance: An organizational learning perspective. *Journal of Product Innovation Management*, 15, pp. 403-422.

Alpert F. (1994), Innovator buying behavior over time: the innovator buying cycle and the cumulative effects of innovations, *Journal of Product Brand Management*, 3 (2), pp. 50-62.

Amico, E.D. (2003). PLM market expected to double by 2008. *Chemical Week*, 165 (7), p. 49.

Anderson, D. (2002). Mass customization: The proactive management of variety. [Internet]. Available: World Wide Web, http://www.build-to-order-consulting.com/mc.htm

Anonymous. (1998). Dell's make-to-order system leaves competitors in the dust. *Manufacturing News*, 5, p. 13.

Anonymous (1). (2001). Mass customization. [Internet]. Available: World Wide Web, http://www.tc2.com/About/AboutMass.htm

Anonymous (2). (2001). MicroPatent Names the Most Innovative Companies of the 20th Century. [Internet]. Available: World Wide Web, http://www.micropat.com/0/new_century9809.html

Anonymous. (2003). So much to do, so little time. *Machine Design*, 75 (3), p. 51.

Angeloni, J. (2002). Contract manufacturing and outsourcing can yield lower overhead and increase yields. *Military & Aerospace Electronics*, August, p. 30.

Armstrong, P. (2002). The costs of activity-based management. *Accounting, Organizations and Society*, 27, pp. 99-120.

Babad, Y. M., & Balachandran, B. V. (1993). Cost Driver Optimization in Activity-Based Costing. *The Accounting Review*, 68(3), pp. 563-575.

Bass, F.M. (1969). A New Product Growth Model for Consumer Durables, *Management Science*, 15, January, pp. 215-227.

Ben-Arieh, D., Qian, L. (2003). Activity-based cost management for design and development stage. *International Journal of Production Economics*, 83, pp. 169-183.

Berson, A., Thearling, K., Smith, S. (1999). *Building data mining applications for CRM*. Emeryville, CA: McGraw-Hill Osbourne Media.

Bonner, J.M., Ruekert, W.R., Walker, O.C. (2002). Upper management control of new product development projects and project performance. *Journal of Product Innovation Management*, 19, pp. 233-245.

Burt, D.N., Norquist W.E., & Anklesaria, J. (1990). *Zero Base Pricing*. Chicago: Probus Publishing.

Calantone, R.J., Di Benedetto, C.A., Schmidt, J.B. (1999). Using the analytic hierarchy process in new product screening. *Journal of Product Innovation Management*, 16, pp. 65-76.

Carr, L.P. & Ittner, C.D. (1992). Measuring the Cost of Ownership. *Cost Management*, Fall, pp. 42-51.

Cavinato, J. L. (1991). Identifying interfirm total cost advantages for supply chain competitiveness. *Journal of Purchasing and Materials Management*, 27(4), pp.10-15.

Cavinato, J. L. (1992). A Total Cost/Value Model for Supply Chain Competitiveness. *Journal of Business Logistics*, 13(2), pp. 285-301.

Chanover, M. (2000). Mass customization - Who? - What Dell, Nike, and others have in store for you. [Internet]. Available: World Wide Web, http://www.core77.com/reactor/mass_customization.html

Chung, S., Kim, G.M. (2003). Performance effects of partnership between manufacturers and suppliers for new product development: the supplier's standpoint. *Research Policy*, 32, pp. 587-603.

Clarke, P., & Bellis-Jones, R. (1996). Activity-based cost management in the management of change. *The TQM Magazine*, 8(2), pp. 43-48.

Cone, E. (2002). The Few, the Proud, the Cost Effective; the Marine Corps got its orders in 1999: reduce the cost of running its 16 major bases by $110 million a year, by 2004. *Baseline*, Oct. 1, 34.

Conceição, P., Hamill, D., Pinheiro, P. (2002). Innovative science and technology commercialization strategies at 3M: A case study. *Journal of Engineering and Technology Management*, 19, pp. 25-38.

Cooper, R.G. (1985). Selecting winning new product projects: Using the NewProd system. *Journal of Product Innovation Management*, 2, pp. 34-44.

Cooper, R. (1988a). The Rise of Activity-Based Costing—Part One: What Is an Activity-Based Cost System? *Journal of Cost Management*, Summer, pp. 45-54.

Cooper, R. (1988b). The Rise of Activity-Based Costing—Part Two: When Do I need an Activity-Based Cost System? *Journal of Cost Management*, Fall, pp. 41-48.

Cooper, R. (1989a). The Rise of Activity-Based Costing—Part Three: How Many Cost Drivers Do You Need, and How Do You Select Them? *Journal of Cost Management*, Winter, pp. 34-46.

Cooper, R. (1989b). The Rise of Activity-Based Costing—Part Four: What Do Activity-Based Cost Systems Look Like? *Journal of Cost Management*, Spring, pp. 38-49.

Cooper, R., & Kaplan, R.S. (1988). Measure Costs Right: Make the Right Decisions. *Harvard Business Review*, September-October, pp. 96-103.

Cooper, R.G. (2001). *Winning the new product: Accelerating the process from idea to launch, 3rd Ed.* Cambridge, MA: Perseus publishing.

Cooper, R.G., Edgett, S. (2002). NPD practices: The dark side of time and time metrics in product innovation. *Vision*, April-May.

Cooper, R.G., Edgett, S.J., Kleinschmidt, E.J. (2002). Optimizing the stage-gate process: What best practice companies are doing?. *Research-Technology Management*, 15.

Cooper, R.G., Kleinschmidt, E.J. (1988). Resource Allocation in the New Product Process. *Industrial Marketing Management*, 17, pp. 249-262.

Crabb, H.C. (1998). *The virtual engineer: 21st century product development.* New York: American Society of Mechanical Engineers.

Crance, J., Castellano, J., & Roehm, H.A (2001). SBC Enhances ABC. *Industrial Management*, 43(6), pp. 27-32.

Crawford, M.C. (1992). The hidden costs of accelerated product development. *Journal of Product Innovation Management*, 9, pp. 188-199.

Crawford, M.C. (1997). *New products management, 5th Edition*. Homewood, IL: Irwin McGraw-Hill), Cusumano, M.A., Selby, R.W. (1995). *Microsoft secrets: How the world's most powerful software company creates technology, shapes markets, and mages people.* New York: Touchstone.

Dahan, E., Hauser, J.R. (2002). The virtual customer. *Journal of Product Innovation Management*, 19, pp. 332-353.

Dooley, K., Subra., A, Anderson J. (2001). Maturity and its impact on new product development project performance. *Research in Engineering Design*, 13, pp. 23-29.

Dooley, K., Subra, A., Anderson J. (2002). *Adoption rates and patterns of best practices in new product development.* [Research report]: Arizona State University.

Dowsett, T., Strydom, J. (2002). Issues & Perspectives. Project risk mitigation: A holistic approach to project risk management. [Internet]. Available: World Wide Web, http://www.ey.com/global/download.nsf/International/ProjectRiskMitigation_AABS_-_June2002/$file/ProjectRiskMitigation_17Jun02.pdf

Drickhamer, D. (2002). A leg up on mass customization. *Industrial Week*, September.

Drucker, P.T. (1994). The age of social transformation. *The Atlantic Monthly*, 274 (5), pp. 53-80.

Drucker, P.T. (1996). The shape of things to come. *Leader to Leader Institute*, 1, pp. 12-18.

Drucker, P.T. (1998). The discipline of innovation. *Leader to Leader Institute*, 5 (4).

Drucker, P.T. (2000). Managing knowledge means managing oneself. *Leader to Leader Institute*, 16, pp. 8-10.

Edmunds, B., Pullin, J. (2003). Integrate to differentiate. *Professional Engineering*, 16 (5), pp. 49-50.

Ehie, I. C. (2001). Determinants of Success in Manufacturing Outsourcing Decisions: A Survey Study. *Production and Inventory Management Journal*, Fall, pp. 31-39.

Ellram, L. (1993). Total Cost of Ownership: elements and implementation. *International Journal of Purchasing Management*, 29(4), pp. 3-12.

Ellram, L. (1994). A Taxonomy of Total Cost of Ownership Models. *Journal of Business Logistics,* 15(1), pp. 171-191.

Ellram, L. M. (1995). Activity-Based Costing and Total Cost of Ownership: A Critical Link. *Journal of Cost Management*, 8(4), pp. 22-30.

Ellram, L.M., & Siferd, S.P. (1993). Purchasing: The Cornerstone of the Total Cost of Ownership Concept. *Journal of Business Logistics.*

Ellram, L.M., & Siferd, S.P. (1998). Total cost of ownership: a key concept in strategic cost management decisions. *Journal of Business Logistics,* 19(1), pp. 55-84.

Estrin, T.L., Kantor, J., & Albers, D. (1994). Is ABC suitable for your company? Management Accounting (US), 75(10), pp. 40-45.

EuroShoe. (2002). *The market for customized footwear in Europe.* [EuroShoe Project Report]: EuroShoe Consortium, Munich/Milan.

Ferrin, B.G., & Plank, R.E (2002). Total Cost of Ownership Models: An Exploratory Study. *The Journa of Supply Chain Management,* 38(3), pp. 18-29.

Filipczak, B. (1997). It takes all kinds: Creativity in the work force. *Training,* 34, pp. 32-39.

Flint, D.J. (2002). Compressing new product success-to-success cycle time: Deep customer value understanding and idea generation. *Industrial Marketing Management,* 31, pp. 305-315.

Fourt, L.A. and Woodlock, J.W. (1960), Early Prediction of Market Success for New Grocery Products, Journal of Marketing, 25, October, pp. 31-38.

Fowler, S.W., King, A.W., Marsh, S.J., Victor, B. (2000). Beyond products: New strategic imperatives for developing competency in dynamic environments. *Journal of Engineering and Technology Management,* 17, pp. 357-377.

Fralix, M. (2001). From mass production to mass customization. *Journal of Textile and Apparel, Technology and Management,* 1 (2).

Garcia, R., Calantone, R. (2002). A critical look at technology innovation topology and innovativeness terminology: A literature review. *Journal of Product Innovation Management,* 19, pp. 110-132.

Garvin, D.A. (1997). *How to build a learning organization.* [Conference]: International Conference on Strategic Manufacturing, Scotland.

Githens, G.D. (2000). *Strategies for balancing the high-wire dynamics of product innovation.* [Annual seminar & symposium report]: Project Management Institute.

Githens, G.D., Peterson, R.J. (2001). *Using risk management in the front end of projects*. [Annual seminar & symposium report]: Project Management Institute.

Goldstein, I.L., Ford, J.K. (2002). *Training in organizations: Needs assessment, development and evaluation. 4th Edition*. Belmont, CA: Wadsworth.

Gooley T. B. (2003). Solving cross-border roadblocks; Step-by-step process mapping helped one shipper speed shipments across the U.S.-Mexico border. *Logistics Management*, March 1, 49.

Gopalakrishnan, S., Bierly, P. (2001). Analyzing innovation adoption using a knowledge-base approach. *Journal of Engineering and Technology Management*, 18, pp. 107-130.

Grieves, M.W. (2003). PLM-Beyond lean manufacturing. *Society of Manufacturing Engineers Journal*, March, p. 23.

Griffin, A. (1997). PDMA Research on new product development practices: Updating trends and benchmarking best practices. *Journal of Product Innovation Management*, 14, pp. 429-458.

Griffin, A., Page, A. (1996). PDMA Success measurement project: Recommended measures for product development success and failure. *Journal of Product Innovation Management*, 13 (4), pp. 478-496.

Gupta, A., Thomas, G. (2001). Organizational learning in a high-tech environment: From theory to practice. *Industrial Management & Data Systems*, 101 (9), pp. 502-507.

Haapaniemi, P. (2002). *Innovation: Closing the implementation gap*. [CEO Survey]: Accenture.

Hamel, G. (2001). Innovation: The new route to new wealth. *Leader to Leader Institute*, 19, pp. 16-21.

Hamel, G. (2003). Innovation as a deep capability. *Leader to Leader Institute*, 27, pp. 19-24.

Haque, B., Pawar, K.S., Barson, R.J. (2000). Analyzing organizational issues in concurrent new product development. *International Journal of Production Economics*, 67, pp. 169-182.

Hauser, J.R. (2001). Metrics thermostat. *Journal of Product Innovation Management*, 18, pp. 134-153.

Heidenberger, K., Schillinger, A., Stummer, C. (2003). Budgeting for research and development: A dynamic financial simulation approach. *Socio-Economic Planning Sciences*, 37, pp. 15-27.

Hitt, M.A., Ireland, R.D., Hoskisson, R.E. (2002). *Strategic management: Competitiveness and globalization*. Albany, NY: South-western College.

Handfield, R. (2002). Reducing Costs Across the Supply Chain. *Optimize*, December, pp. 54-60.

Hultink, E.J., Hart, S., Robben, H., Griffin, A. (2000). Launch decisions and new product success: An empirical comparison of consumer and industrial products. *Journal of Product Innovation Management*, 17, pp. 5-23.

Hyland, P.W., Gieskes, J.F.B., Sloan, T.R. (2001). Occupational clusters as determinants of organizational learning in the product innovative process. *Journal of Workplace Learning*, 13, pp. 198-208.

Jackson, D.W., & Ostrom, L.L. (1980). Life Cycle Costing in Industrial Purchasing. *Journal of Purchasing and Materials Management*, 16(4), pp. 8-12.

Jones, P.H. (2002). When successful products prevent strategic innovations. *Design Management Journal*, 13(2), pp. 30-37.

Johnson, H. T. (1991). Activity-Based Management: Past, Present, and Future. *The Engineering Economist*, 36(3), pp. 219-238.

Jones, T. C., & Dugdale, D. (2002). The ABC bandwagon and the juggernaut of modernity. *Accounting, Organizations and Society*, 27, pp. 121-163.

Joslin, D., Clements, D. (1999). Squeaky wheel optimization. *Journal of Artificial Intelligence Research*, 10, pp. 353-373.

Kalpic, B., Bernus, P. (2002). Business process modeling in industry: The powerful tool in enterprise management. *Computer in Industry*, 47, pp. 299-318.

Karash, R. (1995). Groupware and organizational learning. [Conference]: GroupWare'95 Conference, San Jose.

Katz, D. M. (2002). Activity-Based Costing (ABC). CFO.com, Dec. 31, 2002. Retrieved April 20, 2003, from LexisNexis Academic database.

Kee, R. (1995). Integrating activity-based costing with the theory of constraints to enhance production- related decision-making. *Accounting Horizons*, 9(4), p. 48.

Kee, R.C. (2001). Evaluating The Economics Of Short- And Long-Run Production related Decisions. *Journal of Managerial Issues,* 13(2), p. 139.

Keizer, J.A., Halman, J.I.M., Song, M. (2002). From experience: Applying the risk diagnosing methodology. *Journal of Product Innovation Management*, 19, pp. 213-232.

Kendall, S. (2001). How things change. *CIO Magazine*, August.

Kengpol, A., O'Brien, C. (2001). The development of decision support tool for the selection of advanced technology to achieve rapid product development. *International Journal of Production Economics*, 69, pp. 177-191.

Kim, J., Wilemon, D. (2000). *Accelerating the front end phase in new product development.* [Working paper]: School of Management at Silla University, Korea.

Kock, N.F., McQueen, R.J., Corner, J.L. (1997). The nature of data, information, and knowledge exchanges in business processes. *The Learning Organization*, 4 (2), pp. 70-80.

Koufteros, X.A., Vonderembse, M.A., Doll, W.J. (2002). Integrated product development practices and competitive capabilities: The effects of uncertainty, equivocality, and platform strategy. *Journal of Operations Management*, 20, pp. 331-355.

Kraemer, K.L., Dedrick, J. (2002). Strategic use of Internet and e-commerce: Cisco Systems. *Strategic Information Systems,* 11, pp. 5-29.

Kuczmarski, T.D. (2000). *Managing new products: Using MAP system to accelerate growth.* Chicago, IL: Innovation Press.

Kuczmarski, T.D. K&A study: Winning new product and service practices. *Innovation Mindset*, pp. 2-7.

Leenders, M.A.A.M., Wierenga, B. (2002). The effectiveness of different mechanisms for integrating. *Journal of Product Innovation Management*, 19, pp. 305-317.

Lere, J. C., & Saraph, J. V. (1995). Activity-based costing for purchasing managers' cost and pricing determinations. *International Journal of Purchasing and Materials Management,* 31(4), pp. 25-31.

Letza, S.R., & Gadd, K. (1994). Should Activity-based Costing Be Considered as the Costing Method of Choice for Total Quality Organizations. *The TQM Magazine*, 6(5), pp. 57-63.

Lewis, M.A.. (2001). Success, failure, and organizational competence: A case study of the new product development process. *Journal of Engineering and Technology Management*, 18, pp. 185-206.

Li, Z., Possel-Doelken, F. (Ed.). (2000). *Strategic production networks*. Beijing, P.R. China: Tsinghua University Press.

Little, A.D. (1991). *The Arthur D Little survey on the product innovation process*. Cambridge, MA.

Littlejohn, M., Lofink, C. (2001). IBM Global Services. Corporate learning: Blurring boundaries and breaking barriers. [Internet]. Available: World Wide Web, http://www-1.ibm.com/services/insights/

LoFrumento, T. (2003). How profitable are your customers?. *Optimize*, April, pp. 24-31.

Lui, D.T., Xu, X.W. (2001). A review of web-based product data management. *Computer in Industry*, 44, pp. 251-262.

Lukka, K., & Granlund, M. (2002). The fragmented communication structure within the accounting academia: the case of activity-based costing research genres. *Accounting, Organizations and Society,* 27, pp. 165-190.

Maital S. (1991). The profits of infinite variety. *Across The Board*, October, pp. 7-10.

Mallinger, M. (2000). The Graziadio Business Report. The learning organization in practice: New principles for effective training. [Internet]. Available: World Wide Web, http://gbr.pepperdine.edu/001/learning.html

Maltz, A., & Ellram, L. (1999). Outsourcing Supply Management. *The Journal of Supply Chain Management*, 35(2), pp. 4-17.

Marien, E. J., & Keebler, J. (2002). 6 Stages in Supply-chain Costing. *Traffic World*, Dec. 16, 24.

Marshall, J. (2002). More Complex, More Robust: Activity-based costing systems are harnessing the internet and adding new functionality, giving companies more horsepower than ever. Financial Executive 18(1), pp. 44-45.

May, A., Carter, C. (2001). A case study of virtual team working in the European automotive industry. *International Journal of Industrial Ergonomics*, 27, pp. 171-186.

Mello, A. (2001). Tech Update. Mass customization won't come easy. [Internet]. Available: World Wide Web, http://techupdate.zdnet.com/techupdate/stories/main/0,14179,2833563,00.html

Mikkola, J.H. (2001). *Modularity and interface management.* [Working paper]: Copenhagen Business School.

Milligan, B. (1999). Tracking total cost of ownership proves elusive. *Purchasing,* 127(3), p. 22.

Mishra, B., & Vaysmani, I. (2001). Cost-System Choice and Incentives—Traditional vs. Activity-Based Costing. *Journal of Account Research,* 39(3), pp. 619-641.

Moenaert, R.K., Caeldries, F., Lievens, A., Wauters, E. (2000). Communication flows in international product innovation teams. *Journal of Product Innovation Management*, 17, pp. 360-377.

Monckza, R.M., & Trecha, S.J. (1988). Cost-Based Supplier Performance Evaluation. *Journal of Purchasing and Materials Management*, 24(1), pp. 2-7.

Montoya-Weiss, M.M., O'Driscoll, T.M. (2000). From experience: Applying performance support technology in the fuzzy front end. *Journal of Product Innovation Management*, 17, pp. 143-161.

Morphy, E. (2001). Real-time CRM Industry News from Around the World. After the fall: The future of CRM, Part 1. [Internet]. Available: World Wide Web, http://www.crmdaily.com/perl/story/14602.html

O'Herlihy, N. Newsletter Article. Innovation uncovered. [Internet]. Available: World Wide Web, http://www.prospectus.ie/press/art_innovation_uncovered_long.htm

Olson, E.M., Walker, O.C., Ruekert, R.W., Bonner, J.M. (2001). Patterns of cooperation during NPD among marketing, operations and R&D: Implications for project performance. *Journal of Product Innovation Management*, 18, pp. 258-271.

Orfali, R., Harkey, D., Edwards, J. (1999). *Client/Server survival guide, 3rd Edition.* New York: John Wiley & Sons.

Osborne, R. (2002). New product development: Lesser royals – the customer may be king, but too often he's given less than the royal treatment. *Industry Week*, 251 (3), April, pp. 65-67.

Page, A. (1993). Accessing new product development practices and performance. *Journal of Product Innovation Management*, 10 (4), pp. 273-290.

Park, C., & Kim, G. (1995). An economic evaluation model for advanced manufacturing systems using activity-based costing. *Journal of Manufacturing Systems*, 14(6), p. 439.

Pender, L. (2001). Hard times are the best times. *CIO Magazine*, August.

Peterson, RA and Mahajan, V. (1978), Multi-Product Growth Models, *Research in Marketing*, 1, J. Sheth (editor), JAI Press, Greenwich, Connecticut, pp. 201-232.

Piller, F.T., Moeslein, K. (2002). *From economies of scale toward economies of customer integration. [Working paper]*: Technische Universität München, Germany.

Pine, B.J. (1993). *Mass customization: The new frontier in business competition.* Boston, MA: Harvard Business School Press.

Porter, A.M. (1993). Tying down total cost. *Purchasing*, 115(6), pp. 38-43.

Porter, M.E. (1985). *Competitive advantage: Creating & sustaining superior performance.* New York: The Free Press.

Process Mapping: Adding value to Human Resource Management (2002). *Financial Express*, Sept. 6.

Rafiq, A., & Garg, A. (2002). Better Performance Is as Simple as ABC. *US Banker*, August, 56.

Rampersad, H. (2002). Increasing organizational learning ability based on a knowledge management quick scan. *Journal of Knowledge Management Practice*, October.

Redmond, William H. (2002), Interconnectivity in diffusion of innovations and market competition,
Journal of Business Research, 5816, Article in Press.

Rink, D.R., & Fox, H.W. (2003). Using the Product Life Cycle Concept to Formulate Actionable Purchasing Strategies. *Singapore Management Review,* 25(2), pp. 73-89.

Rogers, Everett M. (1995), *Diffusion of Innovations*, New York: Free Press.

Rogers, E.M. and Shoemaker, F.F. (1971), *Communications of Innovations: A Cross-Cultural Approach*, The Free Press, New York.

Rooney, C.R. Innovating at Internet speed. *Innovation Mindset*, pp. 8-9.

Rosenau, Milton D. (1982), Innovation – Managing the Development of Profitable New Products, Wadsworth, Inc., Belmont, California.

Schwandt, D.R., Marquardt, M.J. (2000). *Organizational learning: From world-class theories to global best practices*. New York: St. Lucie Press.

Senge, P.M. (1). (1990). *The fifth discipline: The art and practice of the learning organization*. New York: Doubleday.

Senge, P.M. (2). (1990). The leader's new work: Building learning organization. *Sloan Management Review*, Fall, pp. 7-23.

Senge, P.M. (1996). *Leading learning organizations: The bold, the powerful, and the invisible*. New York:

Serant, C. (2002). Weak Economy A Drag On OEM Outsourcing Trend. *EBN*, December 2, 2002, 3. Retrieved August 23, 2003, from Business & Company Resource Center database. Jossey Bass, Inc.

Senge, P.M. (1998). The practice of innovation. *Leader to Leader Institute*, 9, pp. 16-22.

Senge, P.M., Carstedt, G. (2001). Innovating our way to the next industrial revolution. *MIT Sloan Management Review*, 42 (2), pp. 24-38.

Sheremata, W.A. (2001). Finding and solving problems in software new product development. *Journal of Product Innovation Management*, 19, pp. 144-158.

Shibly, J.J. (2001). What is Organization Learning. A simple introduction to organization learning. [Internet]. Available: World Wide Web, http://www.systemsprimer.com/what_is_org_learning.htm

Shields, M.D., & Young, S.M. (1991). Managing Product Life Cycle Costs: An Organizational Model. *Journal of Cost Management*, 5 (3), pp. 39-52.

Slater, D., Sullivan, J. (2001). Good ideas, Bad timing. *CIO Magazine*, August.

Smith, P.G. (1990). Fast-cycle product development. *Engineering Management Journal*, 2 (2), pp. 11-16.

Smith, P.G. (1996). Your product development process demands ongoing improvement. *Research- Technology Management*, 39 (2), pp. 37-44.

Smith, P.G. (1). (1998). *Field guide to project management*. New York: John Wiley & Sons.

Smith, P.G. (2). (1998). Make time-to-market technologies a bottom-line issue. *Computer-aided Engineering*, 17(4), p. 78.

Smith, G.S. (1). (1999). From experience: Reaping benefit from speed to market. *Journal of Product Ilnnovation Management*, 16, pp. 222-230.

Smith, P.G. (2). (1999). Managing risk as product development schedules shrink. *Research-Technology Management*, September-October, pp. 25-32.

Smith, P.G. (2000). Focus on profit to reap benefit of speed to market. *Steel in Focus*, Winter 2000, pp. 20- 21.

Smith, P.G., Blanck, E.L. (2002). From experience: Leading dispersed teams. *Journal of Produc Innovation Management*, 19, pp. 294-304.

Smith, P.G., Reinertsen, D.G. (1992). Shortening the product development cycle. *Research-Technology Management*, May-June, pp. 44-49.

Smith, P.G., Reinertsen, D.G. (1998). *Developing products in half the time: New rules, new tools*. New York: John Wiley & Sons.

Souder, Wm.E., Buisson, D., Garrett, T. (1997). Success through customer-driven NPD: A comparison of US and New Zealand small entrepreneurial hi-tech firms. *Journal of Product Innovation Management*, 14, pp. 459-472.

Strategic Directions International Inc. (2003). Instrument Business Outlook. R&D: Adjusted for inflation. [Internet]. Available: World Wide Web, http://global.factiva.com/en/srch/search.asp?NAPC=S, Search for the title of "US R&D spending"

Thearling, K. (1999). White Papers. An overview of data mining techniques. [Internet]. Available: World Wide Web, http://www.thearling.com/text/dmtechniques/dmtechniques.htm

Thearling, K. (2002). White Papers. An introduction to data mining: Discovring hidden value in your data warehouse. [Internet]. Available: World Wide Web, http://www.thearling.com/text/dmwhite/dmwhite.htm

Tidd, J. (Ed.). (2000). *From knowledge management to strategic competence: Measuring technological and organizational innovation*. London: Imperial College Press.

Torbenson, E. (1998). Mass customization: As you like it. *CIO Enterprise*, February.

Tsai, W.-H. (1998). Quality cost measurement under activity-based costing. *The International Journal of Quality & Reliability*, 15(7), p. 719.

Tully, S. (1994). You'll never guess who really makes ... *Fortune*, 130(7), pp.124-128.

Vokurka, R.J., & Lummus, R.R. (2001). At what overhead levels does activity-based costing pay-off? *Production & Inventory Management Journal,* 42(1), pp. 40-49.

Verworn, B., Herstatt, C. (2000). *Managing the fuzzy front end of innovation*. [Doctoral thesis]: Technical University Hamburg-Harburg, Germany.

Vijayan, J. (2003). The data builds the products. *Computer World*, February, pp. 32-34.

Wenger, E. (1996). How to optimize organizational learning. *Healthcare Forum Journal*, Jul-Aug, pp. 22-23.

Whitemore, M., Copulsky, J. (2003). CRM R.I.P.?. *Marketing Magazine*, 108(13), p. 11.

Willaert, S.S.A., Graaf, R., Minderhoud, S. (1998). Collaborative engineering: A case study of concurrent engineering in a wider context. *Journal of Engineering and TechnologyManagement*, 15, pp. 87-109.

Zhang, Y.F., & Fuh, J.Y.H. (1998). A Neural Network Approach for Early Cost Estimation of Packaging Products. *Computers & Industrial Engineering*, 34(2), pp. 433-450.

Zimmerman, F.M., Beal, D. (2002). *Manufacturing works: The vital link between production and prosperity*. Chicago, IL: Dearborn Trade Publishing.

Zsidisin, G.A., & Ellram, L.M. (2001). Activities related to purchasing and supply management involvement in supplier alliances. *International Journal of Physical Distribution & Logistics Management,* 31(9/10), pp. 617-634.

Index